REVISE FOR PRODUCT DESIGN:

Resistant Materials Technology

36161

Barry Lambert
John Halliwell

Consultant: Peter Neal

D0543403

Edexcel
Success through qualifications

Heinemann

Heinemann Educational Publishers
Halley Court, Jordan Hill, Oxford OX2 8EJ
Part of Harcourt Education

Heinemann is a registered trademark of Harcourt Education Limited

© Barry Lambert, John Halliwell, 2004

First published in 2004

08 07 06
10 9 8 7 6 5 4 3

British Library Cataloguing in Publication Data is available
from the British Library on request.

10-digit ISBN: 0 435630 92 X
13-digit ISBN: 978 0 435630 92 8

Typeset by 𝍐 Tek-Art, Croydon, Surrey

Original illustrations © Harcourt Education Limited, 2004

Illustrated by 𝍐 Tek-Art, Croydon, Surrey

Printed and bound in Great Britain by Scotprint

Cover photo: © Powerstock

Picture research by Peter Morris

Acknowledgements
The publishers would like to thank the following authors of *Product Design: Resistant Materials Technology
(2nd edition)* – Lesley Cresswell, Barry Lambert and Alan Goodier – for allowing their work to be used as
reference material for this book.

The authors would like to thank the following for permission to reproduce copyright material:

Edexcel for the examination question on p.103, reproduced by permission of London Qualifications Ltd.

Photographic acknowledgements
The authors and publisher would like to thank the following for permission to reproduce photographs:
(t = top, b = bottom, m = middle, l = left, r = right)

Alfa Robot p.148; Aviation images p.64; Harcourt Education Ltd./Gareth Boden p.57bl;
Harcourt Education Ltd/Peter Morriss p.12, p.16, p.17 (both), p.23, p.26, p.28, p.57ml, p.57br; Nokia
p.57mr; Pro-Lok p.149; Rex Features p.55; Siemens p.112;
Science and Society Photo Library p.57tl, p.57tr, p.61; Techsoft p.48.

Every effort has been made to contact copyright holders of material reproduced in this book. Any omissions
will be rectified in subsequent printings if notice is given to the publishers.

Tel: 01865 888058 www.heinemann.co.uk

Contents

Part 1
Introduction

Introduction

How to use this book

To help you with your revision this book is laid out to correspond to the student textbook, *Advanced Design and Technology for Edexcel, Product Design: Resistant Materials Technology*. Units 1, 3, 4 and 6 are included as these are the units that are assessed by written examination papers rather than by coursework.

The Parts

Part 1 (Introduction) introduces you to revision strategy and to the examination papers. Part 2 Advanced Subsidiary (AS) and Part 3 Advanced GCE (A2) correspond to the units and sections found in the specification and in the student textbook.

The Sections

Each section starts with a short introduction and then is divided into topics. At the end of each section there are sample examination style questions for you to try.

The Topics

'You need to' box

This provides you with a bulleted list of the key learning points that you need for each topic.

Key terms box

These are provided as they are important terms and you need to check you understand each of these terms. They are also shown in bold the first time they are used in the topic.

Key points

These points summarise the learning you need to do and each major heading reflects those provided in the '*You need to*' box at the start of each topic.

Examination questions

Sample examination style questions are provided at the end of each topic so you can see the type of questions you could be asked on each topic.

Acceptable answers

These will help you to understand what is required in your answers. The points you need to make for each mark are shown in bold.

Examiner's tips

Experienced senior examiners have provided examination tips throughout the book to help you understand what is required by the questions and how to give a full answer for each question.

Do you want to improve your grade?

What is revision?

Revision is preparing for an examination and is defined as 'reviewing previously learned material'.

Why revise?

The purpose of revision is to:

- refresh your knowledge and understanding of previously learned material
- improve your ability to recall and apply this knowledge and understanding to the questions in the examination.

How can I improve my grade?

You can improve your grade by revising thoroughly all that you have covered during your course. A copy of the Specification and other useful materials can be obtained from the Edexcel website at www.edexcel.org.uk. Use this to check the areas that you have covered during your lessons. You can also ask your teacher to provide you with copies of past papers from Edexcel. By looking at these materials you will begin to develop a 'feel' for what is expected of you during the examination. It will also give you an impression of the style of questions asked.

Do I need to know about all the content in the Awarding Body Specification?

Yes.

Over a period of years, the awarding body is obliged to cover the whole of the Specification. However, this does not mean that you can predict which elements of the Specification are going to be covered the year you sit the examinations. Sometimes an element might occur two years in a row. It is therefore vital that you make sure that you are familiar with the whole subject area and you prepare fully by revising the whole specification.

How does this book help me to improve my grade?

This book will help you in a number of ways. Firstly, it outlines clearly and logically all the topics that are covered in the Specification. Secondly, it highlights key terms that will be useful to you in your revision and it gives hints and comments that have been written by senior examiners. Added together, these will give you a very good idea of what you really need to concentrate on when doing your revision. It will also give you a clear idea of what is required in the answers.

Revision strategy and explanation

Revising for the examinations

Revising thoroughly for each examination paper is a key part of your Product Design: Resistant Materials Technology course. You need to understand and learn the subject matter for each examination paper and you need to practise by answering sample questions.

As you revise, you may struggle to understand some topics. Make a list of these and use your notes, textbook, this revision book and other resources to research the topic. If you need further assistance, ask your teacher for help as early as possible.

Remember that you should not depend on your teacher alone to provide you with everything

you need. It is important that you familiarise yourself with the subject using your own research and background reading.

Examiner's Tips

- When you revise, make your revision active. Do not just read the book or look at notes – jot main points down and make sketches. These will help reinforce your knowledge.
- The more relevant and appropriate the material that you include in your responses in the examinations, the more marks you are going to get.
- It is important to get your revision strategy correct. Try to make your revision active. It is not a good idea to just read through a textbook or your notes. It is a better idea to make notes as you read through. The act of writing down points will help to reinforce points in your mind and you can use the notes that you make as crib sheets later on.

Remember that there is no substitute for learning the material and that you need to do this as you go along. You need to make sure you give yourself enough time to revise properly before each question paper. This revision book will help you to learn your work and to develop your examination skills.

Further information on the whole course and managing your own learning during the course can be found in *Advanced Design and Technology for Edexcel, Product Design: Resistant Materials Technology*, **Part 1.**

Preparing for the examination papers

To do well in Product Design: Resistant Materials Technology examinations, you need to prepare yourself properly for each written examination. Make sure that you familiarise yourself with the structure of the various papers.

- What type of material is likely to be included in the various examinations?

Table 1 How your work at AS and A2 will be assessed

Level	Course units	Examinations	AS (%)	A2 (%)
AS level	**Unit 1** 'Product Analysis'	$1\frac{1}{2}$ hour examination	30%	15%
	Unit 2 'Coursework'		40%	20%
	Unit 3A 'Materials, Components and Systems' **Unit 3B** Either 'Design and Technology in Society', or 'CAD/CAM', or 'Mechanisms, Energy and Electronics'	Two 45-minute papers sat in a $1\frac{1}{2}$ hour examination	30%	15%
A2 level	**Unit 4A** 'Materials, Components and Systems' **Unit 4B** Either 'Design and Technology in Society', or 'CAD/CAM', or 'Mechanisms, Energy and Electronics'	Two 45-minute papers sat in a $1\frac{1}{2}$ hour examination		15%
	Unit 5 'Coursework'			20%
	Unit 6 'Design Exam'	3 hour examination		15%

- How is the paper laid out?
- How many sections and questions are there?
- How are the marks allocated?
- How much time will you have to answer each question?
- What is the examiner likely to be looking for?
- What equipment will be required for the exam?

The following section will go some way towards helping you to start answering some of these questions.

The question papers

During this course you will be required to sit four externally assessed examination papers. At AS Level you will take Unit 1 and Unit 3, and at A2 Level you will take Unit 4 and Unit 6.

- Unit 1 is a Product Analysis paper.
- Unit 3 and Unit 4 are theory papers divided into two sections: A and B. The Section A questions cover Materials, Components and Systems and the Section B questions cover the optional areas of Design and Technology in Society ,or CAD/CAM, or Mechanisms, Energy and Electronics. Your teachers will make the choice of which optional area you will study. You cannot change your options between AS and A2; you must continue the option you studied during the AS course and continue that option into the A2 course.

- Finally, Unit 6 is a Design paper. This paper is referred to as synoptic. This means that this final unit is structured (in such a way as) to cover elements from both the AS and A2 (courses). For this reason, Unit 6 must always be taken at the end of the course.

Unit 1 (AS) Product Analysis

In this paper you will be given a colour photograph or a number of photographs of a product. You will be required to make a thorough analysis of the given product by responding to a number of questions that will be set out in sections a–g. These sections remain similar year on year and will always follow the same basic format. This involves looking at the writing of specifications, the use of relevant materials, methods of production, manufacturing processes, quality, safety, and the appeal of the product.

This paper will be marked out of a total of 60 marks. Three of those marks are given for the quality of written communication, QWC. The QWC mark is geared not only to the correct use of English but also to the correct use of technical language.

You have $1\frac{1}{2}$ hours to answer this paper.

All parts of this paper are compulsory.

Unit 3 (AS) and Unit 4 (A2) plus option papers

Unit 3 and Unit 4 are theory papers divided into two sections: A and B.

- Section A questions cover Materials, Components and Systems
- Section B questions cover the optional areas of Design and Technology in Society, or CAD/CAM, or Mechanisms, Energy and Electronics.

Your teachers will make the choice of which optional area you will study. You cannot change your options between AS and A2; you must continue the option you studied during the AS course and continue that option into the A2 course.

In Section A, 'Materials, Components and Systems' there will be six questions and will have a total allocation of 30 marks. However, you should note that the marks will not always be divided equally between the questions. For example, some questions may have more than 5 marks, others less depending on the complexity of the question. Section B, the option part of the paper, has two questions each of 15 marks, making a total of 30 marks.

You have $1\frac{1}{2}$ hours to answer both elements of each paper. (It is a good idea to divide the time equally between the two sections, giving 45 minutes per section.)

All parts of these papers are compulsory.

Unit 6 Design (A2 Synoptic)

This examination consists of one compulsory design question. As it is *synoptic*, this paper must be taken at the end of the course and will examine all the areas covered during the two years of your studies.

Pre-release research paper

Approximately six weeks before the examination, you will be given a pre-release research paper. This will give you an idea of the type of product to expect in your examination. This document will outline a context for the examination and will suggest areas that you might consider researching. Take careful note of these points as they are designed to help you. The research that you undertake during the preparation period is very important as you will be expected to refer to it in the examination. The examination itself is referred to as 'open book'. This means that you are able to take into the examination room all the research material that you have undertaken. You can refer to this material throughout the examination. However, there are two things you are *not* permitted to do. Firstly, you are not allowed to pre-draw any material. All the design work that you do must be done on the pre-printed sheets handed out on the day of the examination. Secondly, you may not use ICT during the examination. This includes CAD or Internet research. Any material or research that you have gathered for the examination will not be collected or assessed by the examiner.

Examiner's Tips

- • All the questions in all the papers are compulsory. In order to gain maximum marks, you must attempt the entire question and all parts of that question.
- Before the examination, look at as many past papers as possible and work through them to give you an idea of the type of question to expect. Your teacher should be able to help provide these.
- At the start of each question you may find it useful to jot down some notes about the points that you want to cover in your response. If you do not want these quick notes to be marked by the examiner, just put a line through them and the examiner will ignore them.
- The key word that appears in all the examiners' mark schemes is 'justification'. You need to justify all the points that you have made. It is a good idea to assume that the examiner knows nothing about the subject and that you should write down everything that you feel is relevant to your answer.

The Unit 6 examination

Unlike the other papers in the Design and Technology suite of examinations, where candidates write in examination answer booklets, the Design paper must be answered on the pre-printed, one-sided, A3 sheets provided. You should only use one side of each A3 sheet and should not need to use extra sheets. The design task is printed at the beginning, outlining the exact requirements of the design problem. The pre-printed A3 sheets also include clear statements at the top of each page outlining the requirements of each section together with suggested times. It is very important that you take note of the comments at the top of each sheet together with the number of marks available for each individual section. By doing this you should be able to understand more of what the examiner will be looking for and consequently gain more marks. The best approach to this examination is to think of it in terms of (as) a mini-project rather similar to the coursework projects that you have produced for the course. This examination should contain the elements of research, analysis, the generation of ideas, the developing of solutions, representing and illustrating your final solution, the laying out of production plans and then testing and evaluating your designs.

You have three hours to answer this question.

You must respond to all elements of the paper to gain maximum marks.

Answering the examination questions

To do yourself justice it is very important that you read the questions on the paper. Remember that all the questions must be answered. Make sure that you look carefully at the number of marks allocated to each section of each question. This will give a very good clue as to how many responses or points are required in your answers. There is a very strong likelihood that if four marks are indicated on the paper, the examiner will require four points in your answer. If you do not have the correct number of points, you will not gain the marks.

In many of the questions you will be asked to produce notes and sketches or drawings. If they are needed, make sure that you include *good*

quality, well annotated, clear sketches to back up your written responses. If you are asked for drawings and do not include them, you will not get the marks available and you will not do yourself justice. It is a good idea to adopt a colour coding system in your drawings. For example, you might think of using one colour for plastics, one colour for metals and another for woods. Stick to those colours throughout the paper. Any way of making it easier for examiners to understand what you are representing is a good thing (where the candidates are concerned).

When you write your responses, you need to make sure that the examiner can understand your answer. For example, make sure that your responses are numbered correctly. If you have found it necessary to draw on extra sheets, you should indicate clearly on your pre-printed answer sheets where the rest of the answer can be found.

There are some key words that appear in questions that give a clue to the type of response required. If you see the words 'discuss', 'explain' or 'outline', the examiner is expecting a fairly detailed justification of your answer and not just single word responses.

The main thing to remember is to read the questions very carefully and think before you start to write. The examiner can only give credit for what you have written on the paper and not what he or she thinks you might mean.

Examiner's Tips

• Remember, a good quality, well-annotated drawing can put as much information across to the examiner as a whole page of written notes.
• Read the question thoroughly so that you really understand what the examiner is asking you to do. Do not just skim through it.

Remember, thorough revision and preparation for your exams will give you the confidence to tackle the questions to the best of your ability so that you do as well as you can in each paper.

Part 2
Advanced Subsidiary (AS)

UNIT 1
Industrial and commercial products and practices (R1)

This unit will develop your understanding of industrial and commercial practices through the investigation of products. It will involve the analysis of a range of manufactured products to gain an in-depth understanding of product design, development and manufacture.

This unit will be assessed externally through a $1\frac{1}{2}$ hour Product Analysis examination. You will be given details and illustrations of a product and you will be expected to answer specific questions that relate to the product. You should use appropriate specialist and technical language in the examination and correct spelling, punctuation and grammar.

Basic product specification

You need to

know how to develop product design specifications for a range of products under the following headings:

☐ purpose/function
☐ performance
☐ market
☐ aesthetics/characteristics
☐ quality standards
☐ safety.

KEY TERMS
Check you understand these terms

purpose/function, performance, market, aesthetics/characteristics, quality standards, safety

Further information can be found in *Advanced Design and Technology for Edexcel, Product Design: Resistant Materials Technology*, Unit 1, section a).

KEY POINTS

Purpose/function

The **purpose** or **function** of a product focuses on the aim or end-use of the product and how the product will be used or what it should do.

The statements should reflect what the product does. For example, an electric breadmaker contains a heating element in order to bake the bread; it has an integral mixing bowl; it has several pre-programmed baking/mixing routines. Saying what it does should refer to specific features about the product. For example, it mixes the ingredients, warms the dough gently before baking begins, and sounds an alarm to indicate that the baking is complete.

Performance

The **performance** of a product is related to the materials and components from which it is made and how it should operate.

Market

This falls into two separate categories:

- the requirements of the **market** – must the product be designed in such a way that it meets design/technological trends and developments? For example, a mobile phone may need to be able to take and send pictures
- user requirements and target market groups (TMG) – who has the product been designed for? How big are they? Is it for a child or an adult?

Aesthetics/characteristics

The **aesthetics** or **characteristics** describe how a product looks and feels – its form, texture and colour – and how this can best be exploited to meet the target market. What does the product do for the user's image, e.g. if they have the latest mobile phone? Reference must also be given to how the user interacts with the product – ergonomics.

Quality standards

Quality standards can be achieved in the product through the use of quality control (QC), quality assurance (QA) and total quality management (TQM). All statements must be specific to the product and thoroughly justified

– related to size, dimensions and tolerances. Quality in use is also an essential area that must be commented upon.

Safety

Safety relates to how the product is to be manufactured safely in production and about the safety of the user and product in use. Statements must be specific to the product and processes used in production and should be justified. Although specific British standards knowledge is not required, a reference to the type of standards that might be used should be included. For example, the electric breadmaker should be double insulated and a 'hot' warning light should be visible.

Examiner's Tip

The specification headings will be reproduced on the examination paper. You are expected to address at least four of the specification headings in your answer. In this part of the paper, it is sufficient to answer with bullet points and short sentences. Repeated points or general statements, such as 'lightweight' and 'colourful', may not earn marks unless they are justified, explained or quantified.

EXAMINATION QUESTION

Try answering this yourself.

 Q1 *Outline the product specification for the flask pictured opposite. Address at least **four** of the following headings:*

- *function/purpose*
- *performance*
- *market*
- *aesthetics/characteristics*
- *quality standards*
- *safety.* **(7 marks)**

Fig. 1.1 A flask

Materials and components

You need to

understand how the properties and
working characteristics influence the
choice of materials and components used
in a range of products:

☐ the properties of metals, alloys, plastics
and woods
☐ manufactured boards and composites
☐ standard components.

KEY TERMS

Check you
understand these terms

ferrous, non-ferrous, alloy, thermoplastics, thermosetting
polymers, hardwoods, softwoods, composite

Further information can be found
in *Advanced Design and Technology
for Edexcel, Product Design: Resistant Materials
Technology*, Unit 1, section b).

The whole of this part in the examination will
relate to the justification and use of materials or
components for specific applications within the
selected product.

Materials and components for a particular
product or application can be selected and
justified using these criteria:

• the manufacturing processes involved
• appropriate finishing techniques
• the dimensional accuracy of the finished
product
• cost.

The choice of a material or component will take
into account a combination of these factors.
Sometimes a compromise must be made.

KEY POINTS

The properties of metals and alloys

Ferrous metals are those metals that are
mainly made up from ferrite or iron. They
include small additions of other substances, for
example, mild steel contains between 0.15–0.3
per cent carbon.

Non-ferrous metals are those that contain
no iron, for example, copper, aluminium and
zinc.

An **alloy** is a metal that is formed by mixing
two or more metals together to give a new

Table 1.1 Properties and applications of metals

Material	Properties	Applications/uses
Low carbon steel	Soft, ductile, malleable (see page 31)	Wire, rivets
Mild steel	Ductile (see page 31) and tough	Car body panels, nuts and bolts
Medium carbon steel	Harder than mild steel but less ductile	Springs, axles and garden tools (spades and forks)
Stainless steel	Corrosion resistant	Cutlery, kitchen sinks
Copper	Malleable and ductile. Excellent conductor of heat and electricity	Cables, car radiators, PCBs, electrical cable
Aluminium	Ductile, machines well, good fluidity	Castings, window frames
Zinc	Ductile and easily worked. Excellent corrosion resistance	Die casting alloys, protective coating for steel

metal with improved and enhanced properties. On occasions, other elements are included. Steel and stainless steel are ferrous alloys whereas brass is a non-ferrous alloy.

The properties of plastics

Plastics offer a wide range of manufacturing possibilities. The nature of the material and the processes involved in manufacture mean that products require very little surface finishing. Plastics can be vacuum formed, injection moulded or blow moulded.

Plastics essentially fall into two main groups: **thermoplastics** and **thermosetting polymers**.

The properties of woods

Wood is a natural resource and consequently the quality of the material produced is variable. If it is not handled and treated correctly, wood can warp, twist and cup. It is also prone to decay and fungal attacks. Before the timber can be put to any real use it must first be converted (cut into manageable sections) and then seasoned to reduce the moisture content to an acceptable level.

Woods can be cut, shaped and joined in many ways. They can also be finished in many ways, which enhance the grain and figure. Woods are classified into either **hardwoods** or **softwoods**.

Table 1.2 Characteristics of plastics

Thermoplastics	Thermosetting plastics
• Long chains of molecules tangled together • Few cross links exist between the chains • When heated they become soft, allowing them to be formed • On cooling they become stiff • Can be reheated and reshaped	• Long chains, which are cross linked, resulting in stiff rigid molecular structure • Can only be heated and formed once • On cooling they set in a rigid and permanently stiff molecular structure

Table 1.3 Properties and applications of plastics

Plastic	Properties	Applications/uses
Acrylic	Stiff, hard, durable, easily scratched, good electrical insulator	Baths and bathroom furniture, car indicator covers/reflectors
Polystyrene	Conventional: light, hard, stiff, brittle, low impact strength Toughened: increased impact strength Expanded/foam: buoyant, lightweight, good insulator of sound/heat	Model kits, disposable cups and plates Children's toys Sound and heat insulation, packaging
PVC	Good chemical resistance and weather resistance, stiff, hard, tough, lightweight	Water pipes, guttering, vinyl records, window frames
PET	Very good alcohol and oil barrier. Good chemical resistance, high degree of impact resistance and tensile strength	Drinks bottles
Polypropylene	Light, hard, good impact resistance and chemical resistance. Can be sterilised, good resistance to work-fatigue	Medical equipment, syringes, containers with integral hinges, nets, kitchenware

Table 1.4 Properties and applications of woods

Wood	Type	Properties	Applications/uses
Beech	Hardwood	Hard, tough, finishes well but is prone to warping. Turns well	Workshop benches, kitchen implements, children's toys, furniture
Jelutong	Hardwood	Straight-grained, soft, fine, even texture	Pattern making, carving, drawing boards
Iroko	Hardwood	Durable, heavy, oily, cross-grained	Substitute for teak (furniture) veneers, garden furniture
Pine	Softwood	Straight-grained but knotty, easy to work, cheap and readily available	Construction work, floorboards and roof joists, interior joinery, furniture
Cedar	Softwood	Lightweight, knot free, weak, straight-grained	Timber cladding, sheds, wall panelling

Manufactured boards

Manufactured boards are available in large wide sheets with the most common size being 2440 × 1220 mm. They do not warp and twist in the same way as natural timbers, but if left unsupported, they will sag. Plywood and blockboard both have a natural timber finish on their external surfaces in the form of veneers, thin layers of natural timbers, which create the impression of a solid wood finish.

Composites

Composite material, such as glass reinforced plastic (GRP) or carbon fibre, are made by bonding two or more materials with glues or resins. The composite materials formed have improved mechanical properties.

Both GRP and carbon fibre consist of a reinforcing material in the form of fibrous matting and a bonding agent, or 'matrix', in the form of resins. Composites are generally chosen because they can be formed into very complex shapes. They also have a very good strength-to-weight ratio, that is, they are generally very strong yet comparatively light in weight. They are therefore generally used in structural objects, such as boat hulls, ski poles and lightweight protective body armour.

Standard components

Standard components are those components that are 'bought in' from specialist suppliers. It makes economic sense for manufacturers to buy in such components rather than investing in machinery to make them, especially if they are very specialised. The three categories of standard components are shown in Table 1.6.

Table 1.5 Manufactured boards

Plywood	Blockboard	MDF
Made up from odd number of layers stacked at right angles	Long strips of softwood run along the length of the board	Although prone to water damage, it is very stable. Easily joined using special knock-down fittings
Veneered outside faces	Veneered outside faces	Capable of taking a wide range of finishes
Special waterproof grades, used in the boat-building industry	Solid material used in reproduction furniture and fire doors	Extensively used in flat-packed mass produced furniture

Table 1.6 Standard components

Standard components	Specialised components	Sub-assemblies
Nuts, bolts, washers, nails, screws, rivets, self-tapping screws	Gears, bushes, bearings, cams	Complete printed circuit boards, gearbox assemblies, cam mechanisms

EXAMINATION QUESTIONS

Example questions and answers.

Q1 *The photograph shows a coat peg fixed to a wooden door. It has been formed from an aluminium alloy and has been fixed in place with steel screws.*

Justify the use of:

a) *non-ferrous metal for the coat peg* **(3 marks)**
b) *steel mounting screws.* **(3 marks)**

Acceptable answer

a) The aluminium alloy does **not require any surface finishing**. It is **relatively easy to bend and form** and is **lightweight yet quite strong** in being able to resist the bending forces exerted upon it.
b) The steel mounting screws are **strong and will withstand the torsional forces** as the screw is inserted. They are **widely available in many sizes** and therefore **they are very cheap**.

Fig. 1.2 A coat peg fixed to a wooden door

Working characteristics

You need to

know about the relationship between working characteristics and materials selection related to the product range:

☐ aesthetic properties
☐ physical and mechanical properties.

know about the relationship between finishes, properties and quality related to the product range:

☐ resistance to degradation

☐ surface coating
☐ preservation of materials and the effect of damp and infestations, their consequences on woods and the oxidation of metals
☐ surface decoration.

 KEY TERMS
Check you understand this term

aesthetics

> 📖 **Further information can be found** in *Advanced Design and Technology for Edexcel, Product Design: Resistant Materials Technology*, Unit 1, section b).

KEY POINTS

Aesthetic properties

Aesthetics is primarily concerned with our own perception of the world, environment and all around us. We acquire and develop this understanding through our senses. Similar products, such as kettles, all function and perform in a similar way. However, the two kettles in Figure 1.3 might have very different aesthetic appeal.

The two kettles are made from different materials, using different manufacturing methods, giving different visual appeal. Words such as *soft, hard, cool, smooth, stylish* and *sleek* describe these products. Despite having identical functions, it is the aesthetic appeal of a product that will determine its success or failure on the high street.

Physical and mechanical properties

Every material has specific physical and mechanical properties that make it suitable or not for a specific application. The properties of both metals and plastics change when heat is applied. This makes them suitable for processes such as casting, welding, injection moulding and blow moulding.

Properties of materials can also be improved by making alloys or composites. The steel used for drinks cans and domestic radiator panels is malleable and retains its shape once punched or pressed. GRP speed boats often have their hulls lined with an expanded form of polystyrene because it has excellent buoyancy properties.

Resistance to degradation

Materials, such as bearing surfaces and cutting edges, wear out and wood will decompose

Fig. 1.3 An injection-moulded plastic kettle and a stainless steel whistling kettle

over time. Garden furniture must be treated appropriately with preservatives, such as teak oil, to prolong its life and use. Ferrous metals will oxidise and ultimately rust if surfaces are not treated whereas non-ferrous metals will simply tarnish and suffer surface discolouration. Painting, die coating and blacking are all suitable finishes for ferrous metals. Although plastics take years to rot down, their colour fades when they are left in direct sunlight.

Table 1.7 Types of finish

Finish type	Description
Paints	Surface preparation is essential with most painting. Surfaces should be cleaned and rubbed down before being primed and, if necessary, undercoated. The only exception to this is when using Hammerite or water-based gloss and emulsions. Paints come in endless colours but mainly fall into three basic categories: oil based, polyurethane and emulsions.
Varnishes	A plastic type of finish made from synthetic resins. They generally provide a tough, waterproof finish. They are available in a range of colours and finishes, such as gloss, satin or matt.
Electroplating	Often used to give metals, such as brass and copper, a coating of a more durable and decorative metal, such as chromium or silver. The process is carried out by electrolysis.
Anodising	Although aluminium and its alloys corrode very slowly, they can be anodised to reduce and inhibit corrosion even further. Coloured dyes can also be introduced into the process.
Preservatives	Wooden products left outside must be protected from the wet, fungus decay and attacks from pests. Preservatives are used, which improves their resistance to the wet.

Surface coating/preservation/decoration

Surface coating and decoration offer an extensive range of finishes and treatments that can be applied to woods and metals (Table 1.7).

Surface decoration

Most materials can be given some form of surface decoration by means of engraving, burning or the application of vinyl graphics. Plastics require little surface finishing because of the nature of the material and manufacturing processes involved, such as blow and injection moulding. However, products are often screen-printed with company and product names. This is a fast, highly automated process, which can even be applied over curved surfaces.

> ### Examiner's Tip
>
> In this part of your examination, you will be asked to justify the selection of the materials or components used in the product. You need to refer to the properties/characteristics of the materials or components that make them appropriate for the function they perform or that make them suitable for manufacturing. Your answers must contain sufficient detail to provide adequate justification.

EXAMINATION QUESTIONS

Try answering this yourself.

 Q1 *A flat-pack garden table is:*

- *made from iroko and has been finished with teak oil*
- *fastened together with brass screws.*

Justify the use of:

(a) teak oil as a finish **(3 marks)**
(b) brass screws. **(3 marks)**

Scale of production

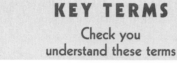

KEY TERMS

Check you
understand these terms

one-off, batch, high volume

📖 Further information can be found in *Advanced Design and Technology for Edexcel, Product Design: Resistant Materials Technology*, Unit 1, section c).

KEY POINTS

Selection of materials and components

The selection of materials and components is determined by the manufacturing processes involved and the scale of production. However, in an attempt to reduce manufacturing costs and overall costs, products often have to be made in large quantities. The manufacture of any product will fall into one of three categories: one-off, batch or high volume.

One-off manufacture is seen in bespoke or commissioned work, such as jewellery, furniture or shop interiors. Despite their size, the Millennium Dome and a football stadium are both examples of one-off design and manufacture. One-off items cost more since a premium has to be paid for the unique features and work involved in both the design and manufacturing elements.

Batch production involves the manufacture of identical components or products in specified, pre-determined 'batches' varying in size from tens up to thousands. If more products are needed after manufacture has stopped, it can be restarted and another batch produced. Tooling, machinery and the workforce are kept as flexible as possible, and careful tool and machinery management are essential. Complicated software programs are used to ensure that schedules and deadlines are adhered to.

High volume production (mass production) has long production runs. It requires high tooling and investment costs and expensive

Table 1.8 Scales of production

One-off: custom-built kitchen	Batch: Olympic medals	High volume: garden table/chairs
• Made to measure • Choice of materials • Choice of finishes • Individual design • Unique • High quality • High cost – very skilled labour • Time consuming • Slow production • Need to use standard components in order to reduce costs	• Same mould used, limited production • Identical quality • Reduced production costs • Limited market • Increased production rates • Same mould/different materials	• Large market • Bulk buying • Quality control • Economies of scale • Rapid production levels • Standardised components • Cheaper to produce • Less skilled labour • Greater level of production planning required • JIT stock control

machinery. Mass production quite often runs non-stop, 24 hours a day, seven days a week. Despite the high costs involved initially, unit costs are often very low, with very little input from skilled operators.

In both batch and high volume production, bulk buying allows materials, components and sub-assemblies to be purchased more cheaply. This reduces the overall cost of the product. Greater use is also made of ICT in the planning of production. This will be used for the drawing up of schedules, quality control, and making sure that materials and other components are ordered and delivered on time, JIT (just in time).

Examiner's Tip

In this part of your examination, you will be asked to give a number of reasons why a particular product has, for example, been mass produced. One-word answers are insufficient – there must be clear evidence of justification for each reason given. Where 4 marks are available for four reasons, each reason for mass production must be carefully justified to score all four marks.

EXAMINATION QUESTION

Example question and answer.

Q1 *A stackable garden chair has been made from polypropylene (PP). Give **four** reasons why the chair is mass produced.* **(4 marks)**

Acceptable answer
The market for this chair is large since it appeals to the home market and the **hospitality market where it would be used in marquees** and similar events.

Since the chair would be injection moulded, it would **be produced quickly in large numbers relatively cheaply** once the **initial set-up costs had been covered**.

Because of the nature of the plastic material, **it requires no finishing** and therefore **less skilled labour is required, further reducing production costs**.

The **quality of the product can be carefully monitored by inspection**, but since the chair is likely to be **made on a single machine, the quality of the product will be** maintained.

Manufacturing processes

You need to

know how one-off, batch and high volume manufacture is achieved using:

☐ preparation
☐ processing
☐ assembly
☐ finishing.

KEY TERMS

Check you
understand these terms

preparation, processing, assembly, finishing

Further information can be found in *Advanced Design and Technology for Edexcel, Product Design: Resistant Materials Technology*, Unit 1, section d).

KEY POINTS

Preparation

The **preparation** stages of manufacturing are related to the selection of materials and the actual manufacture of jigs, patterns, special tools and moulds, *not* to the design and development stages. At this stage, once the moulds have been produced, a number of test

Table 1.9 Processing methods

Process	Main stages	Applications
Die casting	<ul><li>Suitable materials are selected</li><li>Material fed into machine</li><li>Forced into a chamber using hydraulic ram or Archimedean screw</li><li>Molten material is then plunged into mould cavity</li><li>Dwell time to allow for preliminary cooling to start</li><li>Mould opens</li><li>Product is removed using ejector pins</li><li>Mould is cleaned and closes again whereupon the whole process is repeated</li></ul>	Pencil sharpeners, toy cars
Sand casting	<ul><li>Selection of suitable materials</li><li>Mould or pattern packed with sand into cope and drag with sprue pins</li><li>Cope and drag split and mould or pattern removed</li><li>Sprue pins removed and channels opened up</li><li>Cope and drag put back together</li><li>Metal heated to a molten state and poured into empty cavity</li><li>Time allowed for molten metal to solidify</li><li>Cope and drag broken out to reveal cast product</li></ul>	Workshop vices
Injection moulding	<ul><li>Plastic granules fed into hopper unit</li><li>Fed through the heating element by an Archimedean screw thread</li><li>Plasticised material collects in a shot chamber</li><li>The screw thread acting as a hydraulic ram injects the plasticised material into the mould cavity</li><li>The ram maintains the pressure while the plastic solidifies – this is known as the dwell time</li><li>The mould splits and opens</li><li>Ejector pins are used to release the formed product from the open mould</li><li>The mould closes in preparation for the procedure to be repeated</li></ul>	Kettles, kitchenware
Blow moulding	<ul><li>Plastic granules fed into hopper unit</li><li>Fed through the heating element by an Archimedean screw thread</li><li>Plastic parison formed by extrusion</li><li>Split mould opens as the parison is lowered</li><li>Mould closes, trapping parison</li><li>Compressed air blown into parison</li><li>Plastic forced to the side of the mould</li><li>Chilling effect causes the plastic to set</li><li>Formed product drops out when the mould is opened</li></ul>	Bottles, toy ducks
Vacuum forming	<ul><li>The mould is placed on the platen and the table lowered</li><li>The appropriate material is clamped down around its edge to create an airtight seal</li><li>The sheet is heated uniformly until it becomes soft and flexible</li><li>Air is blown into the softened sheet to stretch it uniformly</li><li>The table is raised pushing the mould into the softened sheet</li><li>The air below the trapped sheet is removed</li><li>Atmospheric pressure forces the softened sheet down over the mould</li><li>The material is allowed to cool</li><li>Air is blown back up to help release the formed sheet from the mould</li><li>The formed sheet is removed from the mould</li></ul>	Chocolate box inners, car dashboards, acrylic baths

Table 1.10 Surface coating and finishing

Surface coating	Applied finishes	Self-finishing	Surface decoration
• Dip coating • Paints • Anodising • Timber • Preservatives	• Varnishing • Paints	• Plastics • Injection moulding • Blow moulding • Vacuum forming • Die casting	• Engraving • Etching • Vinyl transfer

products would be made, allowing for numerous quality checks and tests to be carried out prior to full production starting.

Moulds for blow moulding and injection moulding are often split dies; these must include draft angles ,which allows the product to be easily removed once formed. Such moulds are made on Computer Numerically Controlled (CNC) machinery by a combination of milling and spark erosion.

Cast products, whether sand or die cast, also require moulds or patterns. Die casting uses a metal mould with a cavity – this is produced in a similar fashion to the injection and blow moulding dies. With sand casting, a pattern is required of the product/component to be cast. The moulds are made slightly larger than the final finished product to allow for contraction of the metal on cooling and for any surfaces that might need to be machined.

Processing

The **processing** stage is best described as the actual manufacturing of the product or component. The five main manufacturing processes are: die casting, sand casting, injection moulding, blow moulding and vacuum forming.

Assembly

The **assembly** of all the component parts is another critical stage in the production process. A schedule will help the assembly line workers by detailing the order in which the various parts and sub-assemblies need to be assembled. A prime concern of the designer in the early stages of design will have been to take into account the process and order of assembly.

In some cases, push fits are used to fit together various components. In other cases, fastenings, such as nuts, bolts and rivets, are used. Most assembly takes place manually on an assembly line but some assembly is completed automatically by machines. In electrical products, bought-in sub-assemblies, such as electrical leads, sockets, switches and heating elements, would be used. Careful checks and tests are also carried out at this stage to ensure that all parts are fitted correctly and accurately.

Finish

Quality control is an important aspect of the **finishing** stages. It is at this point that the product will undergo any final testing before it is packaged and dispatched. However, prior to this, surface **finishing** will have taken place.

Company logos and labels are prominent features on most products. Company logos are used for advertising and brand identity. Labels, such as Kitemarks or electrical insulation, have to be affixed to the product to indicate that it has passed the relevant and appropriate British and international standards.

Examiner's Tip

In order to score marks in this section, it is important to be specific in your answers. Ask yourself: what steps does the manufacturer have in place to ensure quality? You must identify the nature of the quality check and the procedure under inspection. Because many quality standards checks take place in production, similar answers could appear in both of the quality sections. However, repeating answers used in the previous section will not score marks. Check that you have not used the same answer twice.

EXAMINATION QUESTION

Try answering this yourself.

 Q1 *Describe, using notes and sketches, the stages of manufacture of the watering can under the following headings:*

- *preparation (of tools and equipment)* **(2 marks)**
- *processing* **(10 marks)**
- *assembly* **(2 marks)**
- *finish.* **(2 marks)**

Include reference to industrial manufacturing methods in your answer.

Do NOT include detailed reference to quality control and safety testing in this section.

Fig. 1.4 A plastic watering can

Quality

You need to

understand the importance of quality of design in manufacture including:

- ☐ **quality control in production**
- ☐ **quality standards.**

KEY TERMS

Check you understand these terms

quality assurance, total quality management (TQM), quality control, quality standards

Further information can be found in *Advanced Design and Technology for Edexcel, Product Design: Resistant Materials Technology*, Unit 1, section e).

KEY POINTS

Quality control in production

Quality assurance (QA) is a critical part of any company's performance and success in satisfying their customers' needs and expectations.

QA covers each and every aspect of product development from the design stages to the delivery of the product. When manufacturing in high volumes, QA will ensure that the products are manufactured to budget, time and specification. This level and standard of manufacture demonstrates the use of **total quality management (TQM)**. Companies that reach these standards can apply for ISO 9000, the international standard of quality, which enhances their reputation and stature within the manufacturing industry. Features of TQM include the use of jigs, patterns and gauges as a means of carrying out **quality control**. Quality control (QC) has a much more practical emphasis in the way it operates. At critical points in the production process, parts will be measured and inspected against tolerances and specifications, such as those concerning materials, finishes and assembly.

Quality standards

Quality standards are established to ensure that manufacturers meet the requirements and demands of both customers and users. Quality standards have evolved over time and will continue to evolve as technologies improve and consumers demand more. Quality standards are determined largely by organisations such as BSI (British Standards Institute) and ISO (International Organisation for Standardisation). These standards are incorporated into product specifications to serve and protect both the manufacturer and the consumer.

Testing of a product may cover the product's overall performance, its life expectancy and whether it conforms to the standards laid down by BSI or ISO. Products that perform satisfactorily in tests are labelled appropriately.

The Kitemark symbol is awarded to a product that has undergone specific tests to ensure it conforms to the QS laid down in its original specification. This will confirm that the product is safe and reliable to use.

Examiner's Tip

In this section on quality, you will be expected to answer two separate sections: quality control in production and quality standards. In each section, you will need to make a number of points associated with each area. You must justify each point in relation to the specific product being analysed.

EXAMINATION QUESTIONS

Example questions and answers.

Q1 *Discuss the quality issues of the fruit juice bottle on page 28 under the following headings:*

a) *quality control in production* **(4 marks)**
b) *quality standards.* **(4 marks)**

Acceptable answer

a) *Quality control in production*
First the company must ensure the **importance of good design practice when designing the moulds** in which the bottles will be blow moulded. This means the need to ensure that the product can be easily manufactured; **a number of test products should** be made to ensure that the production runs smoothly and that they meet the standards laid down in the initial specification. During manufacture, **tooling and equipment must be consistently monitored in order to maintain standards.** All of these aspects would contribute to **TQM, total quality**

management, for the overall company and product manufacture. Eventually, ISO 9000 would demonstrate to customers and clients that the manufacturers were very good.

b) *Quality standards*
Naturally, the product would be **tested against a pre-determined set of criteria to ensure its overall fitness for purpose** and that it was safe for consumers to use. Once it has passed these tests, it should be **labelled appropriately** to show that it has successfully passed the tests. Tests would also be carried out on the **appropriateness of material selection** to make sure it will last in usual conditions, e.g. being squeezed to get the drink out. It would also be **tested to ensure that it is waterproof**. Guarantees and customer satisfaction surveys would also be used as a means to identify and record that the product conforms to all quality standards.

Health and safety

You need to

understand the principles of health and safety legislation and good manufacturing practice including:

☐ the safe use of the product
☐ safety procedures in production.

KEY TERMS

Check you
understand these terms

legislation, health and safety, risk assessment

Further information can be found in *Advanced Design and Technology for Edexcel, Product Design: Resistant Materials Technology*, Unit 1, section f).

KEY POINTS

Safe use of the product

In today's society and marketplace, unless a product is safe it will not make it to the high street. All products must be safe and reliable in use and therefore must be tested to the appropriate British and international standards. Part of the testing procedure will be against the initial product specification. For a food processor, this will ensure that it conforms to performance specifications, such as motor speeds, mixing/blending capacities and motor power.

Obviously, electrical safety is of prime concern, especially if mains electricity is being used. All electrical products must be tested and passed to ensure they are safe to use and labelled appropriately. The Kitemark symbol, CE mark and double insulation marks for electrical products are all prominently displayed on or moulded into the product casing.

Instructions for use are carefully considered by the manufacturers. They are always included with the product, and guarantees and warranties are provided as an indication of product reliability.

Safety procedures in production

The manufacturer not only has to ensure that the product is safe in use, but also that safety procedures are strictly adhered to in its production. **Legislation** exists and guidance is given on reducing the risk of accidents. This legislation includes:

- The Factories Act 1961
- The Health and Safety at Work Act 1974
- regulations and codes of practices such as COSHH (Control of Substances Hazardous to Health)
- The Management of Health and Safety at Work Regulations 1992.

The **Health and Safety** Executive was created as a regulatory body to ensure that companies complied with the various acts and legislation. They were also given the power to enter and inspect premises in relation to health and safety matters.

One aim of the 1992 Management Act was to ensure that companies and manufacturers carry out assessments to identify potential hazards and risks that could cause harm and damage to workers. This process is called **'risk assessment'**. Hazards are deemed to be anything that has the potential to cause harm or damage. The risk of any hazard must be assessed and evaluated through a process that analyses the chance of injury and the potential degree of damage that could occur. If the risk is deemed to be dangerous above a certain level, it must be eliminated or controlled. The investigation and analysis of such hazards in the production of a product would take into account some of these aspects:

- appropriate health and safety regulations and training of staff using machines and handling tools and materials

- the guarding of machines, especially those that involve human intervention or control
- appropriate ventilation, heating, light and noise levels
- the servicing of machines and log books kept up to date
- appropriate health and safety notices around the factory and machines
- the correct use of personal protective equipment
- that risk assessments have been carried out

- any COSHH data is provided, along with appropriate first aid instructions.

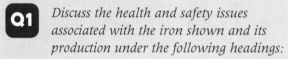

Examiner's Tip

When you are discussing safety in the workplace, do not be tempted to write a long list of clothes: apron, safety shoes, gloves, etc. You will only be given 1 mark for mentioning that you must be aware of the clothing worn in the workshop.

EXAMINATION QUESTIONS

Example questions and answers.

Q1 *Discuss the health and safety issues associated with the iron shown and its production under the following headings:*

a) *safe use of the iron* **(4 marks)**

b) *safety procedures in the production of the iron.* **(4 marks)**

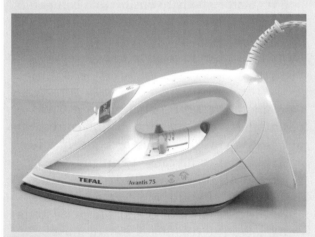

Fig. 1.5 An iron

Acceptable answer

a) **Guidelines and instructions for use are provided** for the user and these must be consulted before using the product. Since the iron has a **water container, it must be electrically double insulated** to ensure that it is and would remain electrically safe if the cable became loose or was burnt through. The iron itself and the **process of ironing is potentially very dangerous**. The user or the clothes being ironed can be burnt and the flex can be damaged if it comes into contact with the hot iron plate/sole for too long. To overcome this, the cable often has a **heatproof protective sleeve covering it**.

b) Injection moulding is involved in the production of the iron and great care should be taken by operatives when using such machinery. Naturally, the machines will be **appropriately guarded**. The **operatives should be wearing the correct personal protective** equipment, such as goggles and ear defenders, because they will be subjected to these dangers for long periods of time.

Risk assessments will also have been completed for the pieces of equipment and machines used by the workers. This will have identified any potential risks and then the potential dangers can be eliminated.

Other health and safety regulations, such as **COSHH, also have to be considered**. This will ensure that all potentially dangerous chemicals, fumes and substances are carefully controlled, handled and stored so that the workplace remains safe for all employees.

Product appeal

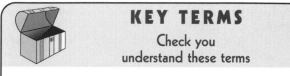

KEY TERMS

Check you understand these terms

form, function, trends, styles, cultural

📖 **Further information can be found** in *Advanced Design and Technology for Edexcel, Product Design: Resistant Materials Technology,* Unit 1, section g).

KEY POINTS

Influences on the design, production and sale of products

One important aspect of product design and development is keeping up with ever-changing trends and styles. There is also a continuing need to embrace and incorporate new and developing technologies.

Over recent years, products such as mobile phones have become smaller, incorporate new technologies and possess higher processing and data handling capacity. The new designs not only embrace all these features but are also more stylish, with enhanced aesthetic appeal. The success of a company is to a large extent dependent upon its ability to develop new products aimed at its TMG (target market group). Manufacturers wishing to maintain and grow their market share must recognise and act upon cultural differences around the world and greater environmental awareness and concern.

In your examination, you will need to discuss the **form**, **function**, **trends/styles** and **cultural** aspects of the appeal of a product.

Form and function – must relate to the product's design features and technical performance.

- How does the product meet the TMG's requirements?
- Who are the TMG?
- What makes the product suitable for the TMG?
- What are the specific characteristics that make it more appealing than other similar products on the market?

Trends and styles – must relate to the aesthetic appeal of the product.

- How does the style of the product appeal to the TMG?
- How does it fit in and enhance the lifestyle of the TMG?
- Does the colour/style meet with current market trends?

Cultural – must relate to and reflect values issues that influence the design of the product.

- Is the product recyclable or has it been made from recycled materials?
- Does it have a Kitemark or other form of guarantee or warranty?
- Where will it be sold?
- Does it represent value for money?
- Is it a cheap quality or high quality item?

Examiner's Tip

You will be asked to discuss the appeal of a product under some or all of the following headings: form, function, trend/styles and cultural. Obviously not every product being analysed will conform to all of the headings, so only two or three headings may be given and the marks allocated accordingly. Points made must be fully explained and justified in order to achieve full marks.

EXAMINATION QUESTION

Example question and answer.

 Q1 *Discuss the appeal of the fruit juice bottle on page 28 under the following headings:*

- *form*   **(2 marks)**
- *function*   **(2 marks)**
- *trends/styles*   **(2 marks)**
- *cultural.*   **(2 marks)**

Acceptable answer

Form – the bottle is an **attractive shape** and **ergonomically designed, making it easy to hold and squeeze**. It is **attractively finished**, partly due to the nature of the materials and processes involved, and it could easily be printed with logos.

Function – the bottle is **easy to fill** since it **can fit under a tap**. The **stopper can easily be opened and closed**, keeping the drink safely inside and **safe to use**.

Trends/styles – the bottle is **modern in shape and will appeal to young cyclists**. Since it is likely to be manufactured in a range of colours, **it allows the individual to select a colour that is best suited to their needs**.

Cultural – as a modern-day cycling accessory, **it will fit into the biking scene, especially for mountain bikes** and riders on long-distance journeys.

> ### *Examiner's Tip*
>
> When you are revising for the examination, choose a product and look carefully at the following areas: product specification, materials, level of production, stages of manufacture, quality, health and safety, and aesthetic qualities.

PRACTICE EXAMINATION STYLE QUESTION

Figure 1.6 shows a photograph of a fruit juice drinks bottle.

The body of the bottle is:

- manufactured from PET
- mass produced in a range of colours to match the flavour of the content.

The lid is:

- mass produced by injection moulding
- tamperproof.

a) Outline the product specification for the illustrated fruit juice bottle. Address at least **four** of the following headings:
  - function/purpose
  - performance
  - market
  - aesthetics/characteristics
  - quality standards
  - safety.   **(7)**

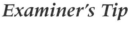

Fig. 1.6 A fruit juice drinks bottle

b) Justify the use of:
- PET in the fruit juice bottle **(3)**
- a tamperproof seal on the lid. **(3)**

c) Give **four** reasons why the fruit juice bottle is mass produced. **(4)**

d) Describe, using notes and sketches, the stages of manufacture of the body of the fruit juice bottle under the following headings:
- preparation (of tools and equipment) **(2)**
- processing **(12)**
- assembly **(1)**
- finish. **(1)**

Include references to industrial manufacturing methods in your answer.

e) Discuss the quality of the fruit juice bottle under the following headings:
- quality control in production **(4)**
- quality standards. **(4)**

f) Discuss the health and safety issues associated with the fruit juice bottle and its production under the following headings:
- safety of use of the fruit juice bottle **(4)**
- safety procedures in the production of the fruit juice bottle. **(4)**

g) Discuss the appeal of the fruit juice bottle under the following headings:
- form **(2)**
- function **(2)**
- trends/style **(2)**
- cultural. **(2)**

h) Quality of written communication. **(3)**

Total for this question paper: 60 marks

Note: Quality of written communication

Three marks are awarded overall for the quality of written communication. In order to achieve the full marks, your written work must include appropriate specialist vocabulary and excellent spelling, punctuation and grammar to communicate consistently with clarity, relevance and coherence. The quality of written communication also includes the correct and appropriate use of technical language.

Unit 3 is divided into two sections:

- Section A: Materials, components and systems (compulsory for all candidates)
- Section B: Consists of three options (of which you will study only **one**).

Section A will be assessed during the $1\frac{1}{2}$ hour, Unit 3 examination. You should spend half of your time answering all of the questions for this section. It is important to use appropriate specialist and technical language in the examination, along with accurate spelling, punctuation and grammar. Where appropriate, you should include clear, annotated sketches to explain your answer.

Ferrous metals, non-ferrous metals and alloys

You need to

have an understanding of:

- ☐ the source and classification of metals
- ☐ the micro-structure of metals
- ☐ common alloys, working properties and common uses of metals.

Further information can be found in *Advanced Design and Technology for Edexcel, Product Design: Resistant Materials Technology*, Unit 3A, section 1.

KEY TERMS

Check you understand these terms

ferrous, non-ferrous, alloys, ductile, malleable, crystals, lattice

KEY POINTS

The source and classification of metals

Metals are divided into three basic categories, as shown in Table 3.1.

With the exception of gold, all metals are found in the form of oxides or sulphates. They need to

Table 3.1 Classification of metals

Ferrous metals	Metals that contain mainly ferrite or iron. Some have small additions of other substances – e.g. mild steel, cast iron. Almost all are magnetic.
Non-ferrous metals	Metals that contain no iron – e.g. copper, aluminium and lead. These are often **ductile**, i.e. can be drawn into thinner sections, and **malleable**, i.e. can be deformed without any cracking or fracturing.
Alloys	Metals that are formed by mixing two or more metals and, on occasions, other elements to improve properties. They are grouped into ferrous and non-ferrous alloys.

Table 3.2 The micro-structure and properties of metals

Structure	Properties	Examples
Close-packed hexagonal (CPH)	Weak, poor strength-to-weight ratio	Zinc, magnesium
Face-centred cubic (FCC)	Ductile, good electrical conductor	Gold, copper, silver, aluminium
Body-centred cubic (BCC)	Hard, tough	Chromium, tungsten

be extracted from the ore before any useful processing can take place.

Ferrous metals are extensively produced in blast furnaces with the addition of carbon, in the form of coke, and limestone, and mild steel with other materials, such as chromium and tungsten.

Non-ferrous metals are mainly produced by electrolysis. This is very expensive because of the electrical energy costs involved.

The micro-structure of metals

With the exception of mercury, all metals are solid at room temperature. In their molten state, they are weak and flow easily. As they cool and solidify, small seed **crystals** are formed and the atoms arrange themselves in a regular pattern – a **lattice** structure. Finally, small crystals and grains are formed. These crystalline structures fall into one of three basic forms and will determine the properties of the metal.

Table 3.3 Properties of some common alloys and metals

Material	Melting point (°C)	Composition	Properties	Uses
Steels	1400	Alloy of carbon and iron		
Low carbon		<0.15% carbon	Soft, ductile	Wire, rivets, cold pressings
Mild		0.15–0.3% carbon	Ductile, tough	Car bodies, nuts and bolts
Medium carbon		0.3–0.7% carbon	Harder than mild steel, less ductile	Springs, axles
High carbon		0.7–1.4% carbon	Hardness can be improved by heat treatment	Hammers, cutting tools and files
Stainless		Medium carbon steel + chromium + nickel	Corrosion resistant	Kitchen sinks, cutlery
Aluminium	660	Pure metal	Malleable, ductile. Good conductor of heat and electricity	Castings, alloys used extensively in the aircraft industry
Copper	1083	Pure metal	Malleable, ductile. Excellent conductor of heat and electricity	Electrical cable, plumbing pipes
Brass	927	65% copper + 35% zinc	Corrosion resistance, casts well	Marine and plumbing fittings
Lead	327	Pure metal	Soft and malleable. Excellent corrosion resistance	Roof covering and flashing, protection against radiation and X-rays

Common alloys, working properties and common uses of metals

An alloy is a metal made by combining two or more metals, and sometimes non-metallic elements, to improve the properties of the original base metals. Table 3.3 on the previous page shows the properties of some common alloys and metals.

Thermoplastic and setting polymers

You need to

☐ **know about the chemical make-up and polymer structure of plastics**
☐ **know about the basic methods of working.**

KEY TERMS
Check you understand these terms

thermoplastic, thermosetting plastics, cracked, Van der Waals, plasticity, polymerisation, covalent bonding

Further information can be found in *Advanced Design and Technology for Edexcel, Product Design: Resistant Materials Technology*, **Unit 3A, section 1.**

KEY POINTS

The chemical make-up and polymer structure of plastics

Thermoplastic and **thermosetting plastics** are a wide and diverse range of substances that exist in both natural and synthetic forms. The main raw material used in the production of plastics is crude oil, consisting of hydrogen and carbon in the form of hydrocarbon naphtha. This is further processed and **cracked**, or broken down, into fragments to form ethylene and propylene.

Thermoplastics

- Made up from long tangled chains of molecules with small cross links.
- The polymer chains are held together by mutual attraction, known as **Van der Waals** forces.
- The introduction of heat weakens the bonds and the material becomes pliable and easier to mould and form. When the heat is removed, the chains reposition and the material becomes stiff once again.
- Acetate, polypropylene and polystyrene are all thermoplastics.

Thermosetting plastics

- Thermosetting plastics set with heat and thereafter have little **plasticity**, the ability to be changed permanently without cracking or breaking. During **polymerisation** (the formation of large chains of molecules), the molecules link side to side and end to end. This is known as **covalent bonding** and gives a very stiff and rigid micro-structure.

- Once formed, they cannot be reheated.
- Polyester resin, which is used in fibreglass, is a thermosetting plastic.

Methods of working

Glass reinforced plastic (GRP)

Glass reinforced plastic is often referred to as fibreglass. It consists of strands of glass fibres that are set rigid in a polyester resin. The polyester resin is initially in the form of a liquid, but when a hardener is added, along with any pigmentation, it will set rigid. The glass fibre strands provide the basic strength, while the resin with its additives bonds the fibres together and provides a very smooth finish. GRP work requires a mould or former over which to form the product. The surface over which the mould is being made must have a high-quality finish. GRP is used in a vast range of products that need its great strength-to-weight ratio, for example, sailing boat hulls and canoes.

Injection moulding

Injection moulding is probably the most versatile process used for commercially moulded plastic products – from bowls and buckets to television casings and dustbins. It is only appropriate for high volume production since the initial cost of the machine and moulds is very high. The unit costs, however, are very low. The process is quite simple; the material is heated into a plastic state before it is injected under high pressure into an enclosed mould. Once cooled, the mould is automatically opened and removed with ejector pins. No surface finishing is required other than perhaps to remove the sprue pins and gates, which are often recycled back into the process.

Blow moulding

Blow moulding is used to form hollow products and components, such as bottles and containers. A hollow length of plastic known as a *parison* is formed and lowered into an open split mould. As the mould is closed, a compressed injector is raised and blows air into the parison. The plastic is forced to the sides of the mould cavity and chills. As the mould opens, the formed product drops out and another parison is lowered into position. Blow moulding is another highly automated process, which produces little waste and only requires any flashing to be trimmed. The machines and mould costs are high and so it is only suited to high volume production.

Vacuum forming

Vacuum forming is used to produce simple shapes from thermoplastic sheet. The most commonly used plastics include high density polystyrene, ABS and some grades of PVC. First a mould of the finished component must be produced. It needs slightly tapered sides to ease the removal of the formed component. A trapped sheet is heated and softened before the mould is raised into it. A pump is then used to expel the air below it and atmospheric pressure forces the softened material over the mould. Vacuum forming is used to produce acrylic baths and plastic packaging, such as that around Easter eggs.

Fusion

Plastic sheets can be fusion welded together using ultrasound. Tiny vibrations cause a rise in temperature at a localised point and, with the application of pressure, the two separate pieces become fused. Fusion welding can also be used to join solid sections. The joint is heated with a hot air torch and a filler rod of the same material is added. This process is useful where adhesives cannot be used because chemical attack would render the joint useless.

EXAMINATION QUESTIONS

Example questions and answers.

Q1
a) *Explain what is meant by the term 'polymerisation'.* **(1 mark)**
b) *Describe the difference in structure between a thermoplastic and a thermosetting plastic.* **(4 marks)**

Acceptable answers

a) Polymerisation is the **formation of large chains of molecules where the molecules link side to side and end to end.**

b) Thermoplastics **are made up from long tangled chains of molecules with small cross links. The chains** are held together by **mutual attraction, known as Van der Waals forces,** and tend to be rather **flexible**.

Thermosetting **plastics set with heat** and thereafter have **little plasticity since the molecules link side to side and end to end.** This is known as **covalent bonding** and results in a **very stiff and rigid micro-structure.**

Hardwoods and softwoods and manufactured boards

You need to

☐ **know the sources and classification of timbers**
☐ **know the characteristics and faults of woods**
☐ **know about conversion and seasoning**
☐ **know about manufactured boards.**

KEY TERMS

Check you understand these terms

softwood, hardwoods, heartwood, sapwood, annual rings, fibres, warping, bowing, twisting, felled, conversion, seasoning

Further information can be found in *Advanced Design and Technology for Edexcel, Product Design: Resistant Materials Technology,* **Unit 3A, section 1.**

KEY POINTS

Sources and classification of timbers

Most commercially grown **softwood** comes from the northern hemisphere, particularly the colder regions, such as Scandinavia. Conifers, often called evergreens, are relatively fast growing and produce straight trunks that make for economic cultivation with little waste. Careful forest management ensures that the worldwide demand for good quality softwoods can be met.

Hardwoods exist in many thousands of species and are grown across the whole world. Most broadleafed trees are deciduous, that is, lose their leaves in the winter (with a few exceptions, such as holly and laurel). Hardwoods are generally more durable than softwoods and offer much more variety of colour, texture and figure. Since they take relatively longer to grow, they are generally more expensive than softwoods. The most expensive and exotic are turned into veneers.

A tree consists of two major parts, an inner '**heartwood**', which gives rise to strength and rigidity, and the outer layers or '**sapwood**', which is where the growth occurs. Growth is seasonal and results in **annual rings**, each indicating one year's growth. As the tree grows, the wood tissue grows in the form of long tube-like cells that vary in shape and size. These are known as **fibres** and are arranged roughly parallel to the trunk, giving rise to the general

Table 3.4 Faults of woods

Warping	A cupping across the width of the board	
Bowing	Occurs along the length of the timber	
Twisting	A twist from side to side along the length of the timber	

grain direction. It is this variation in cell size and make-up that leads to the botanical distinction between hardwoods and softwoods.

Characteristics and faults of wood

Wood is a naturally occurring material and as such produces variable quality. It can be cut and shaped in numerous ways, but the irregularity of grain, knots, warping and twisting are all disadvantages. Wood is also prone to biological attack from insects and fungi. Faults and defects (movement) affect the overall strength and durability as well as marring the visual appearance. Defects can be caused by a variety of factors. For example, shrinkage affects the shape of the board once cut. Movement, which cannot be eliminated since it results from any change in moisture and humidity, exists in three main forms (Table 3.4).

Conversion and seasoning

Conversion

Once a tree has reached full maturity, it is cut down or **felled**. **Conversion** is the process of sawing logs into commercially viable timber. The timber is cut in one of two ways, which are reflected in the price and ultimately the figure and dimensional stability of the wood.

Slab, plain or through and through conversion is the quickest, simplest and cheapest method. A series of parallel cuts are made through the length of the timber, resulting in slabs. The thickness of the slabs can be varied as the log is cut. This method is frequently used on softwoods where the logs tend not to be large in diameter.

Quarter or radial sawn conversion is much more time consuming, hence it is much more expensive. However, the timber produced is much more stable. Quarter sawing results in the grain's figure being exposed, and this is noticeable in oak.

Seasoning

The **seasoning** of wood is the process of removing the excess water and much of the bound moisture from the cell walls. Generally,

Table 3.5 Methods of seasoning

Natural-air seasoning	Kiln seasoning
• Slabs are stacked and air is allowed to flow around them • On average, it takes one year to season 25 mm of thickness	• Steam is used in a kiln to heat the timber • A vacuum is then created in a controlled environment • Very precise moisture content levels can be achieved • Many bugs and fungi are killed

the moisture content is reduced to less than twenty per cent, making it more immune to rot and decay, less corrosive to metals and improving the overall strength and dimensional stability. Seasoning can be carried out in two different ways (Table 3.5 on page 36).

Manufactured boards

The rapid growth in manufactured boards has to a certain extent reduced the demand for prime natural quality timbers. They should, however, not be regarded as a cheap substitute and they present many of their own problems.

Table 3.6 Advantages and disadvantages of manufactured boards

Advantages	Disadvantages
• Large standard-sized sheets • Uniform thickness • Stable in most atmospheric conditions • Grained boards have good strength-to-weight ratio • Thin sheets can be bent easily	• Difficult to join • Exposed edges often need to be concealed • Thin sheets become easily distorted unless held in a frame • Adhesives can blunt cutting tools quickly

EXAMINATION QUESTIONS

Example questions and answers.

Q1
a) *Plain, slab or through and through is one method of converting timber. Give the name of the other method.* **(1 mark)**
b) *Give two disadvantages of plain, slab or through and through converting.* **(2 marks)**
c) *Explain one advantage of kiln seasoning.* **(2 marks)**

Acceptable answers
a) **Radial sawing**.
b) Although it is mainly softwoods that tend to be plain cut, **the figure of the timber is not always fully exposed**. Secondly, since the timber is cut across the annual rings, it does **tend to cup across the width of the board cut**.
c) Kiln seasoning kills **many bugs and fungi** during the process, which means that the **timber is free from both types of infestation**.

Composites, synthetics and manufactured materials

You need to

☐ **know about the manufacture of composite materials.**

KEY TERMS
Check you understand these terms

composite, matrix, carbon fibre, GRP, MDF

Further information can be found in *Advanced Design and Technology for Edexcel, Product Design: Resistant Materials Technology*, Unit 3A, section 1.

KEY POINTS

The manufacture of composite materials

A **composite** is formed when two or more materials are combined by bonding. The new material will have better mechanical and other properties than the materials from which it was formed. Most composites have excellent strength-to-weight ratios, that is, they are stronger than other materials of the same weight or mass.

Composites consist of a reinforcing material that provides the strength and a bonding agent, called the '**matrix**', in the form of glues or resins.

Table 3.7 Composite materials

Composite	Application
Carbon fibre	Carbon fibres are very strong. They are used in many structural components ranging from propeller and rotor blades to body armour and sports equipment, such as golf clubs and skis. Using more carbon fibres in the aviation industry has improved fuel economy since they are very light relative to their strength.
GRPs (glass reinforced plastics)	Glass fibre strands are held in a matrix of polyester resin. The strands are available in a wide range of forms and can be obtained in large mats or loose chopped strands. GRP is best suited to large structural items, such as boat hulls, pond liners and septic tanks.
MDF (medium density fibreboard)	MDF is now widely used in industry and schools. The fibrous element comes from wood waste that has been reduced to its basic fibrous form to produce a homogeneous material. The fibres are then bonded together with a synthetic resin adhesive to produce a uniform structure and fine textured material. This material is widely used in flat-pack furniture, shop fitting and model making. It is capable of taking a wide range of surface finishes, from veneers to paints and foil-backed plastic laminates.

There are some dangers associated with the use of most composites. When cutting and sanding composites, very fine particles of both the fibrous material and bonding resins are given off. Breathing equipment, such as masks, should be worn at all times since the dust can cause severe respiratory problems.

Examiner's Tip

Remember that you will need to revise the whole specification. It is quite possible for a topic to be repeated for a second year and/or to be omitted for a couple of years.

EXAMINATION QUESTION

Example question and answer.

Q1 *Explain, giving an example, what is meant by the term 'strength-to-weight' ratio when applied to composites.* **(5 marks)**

Acceptable answer

The term 'strength-to-weight' ratio applies to the product's **overall strength in relation to its weight or mass. For example, an old-fashioned ski pole might have been made from an aluminium alloy. A newer ski pole made from carbon fibre will be stronger than the aluminium alloy, particularly in withstanding bending forces, but it will be lighter in weight; hence the term 'strength-to-weight' ratio.**

Ceramics, glass and concrete

You need to

☐ know about the sources and manufacture of ceramics, glass and concrete and their industrial applications.

 Further information can be found in *Advanced Design and Technology for Edexcel, Product Design: Resistant Materials Technology*, **Unit 3A, section 1.**

 KEY TERMS
Check you understand these terms

ceramics, glass, concrete, aggregate

KEY POINTS

The sources and manufacture of ceramics, glass and concrete and their inductrial applications

Table 3.8 Ceramics, glass and concrete

Material	Composition	Properties	Applications
Ceramics	Clay, sand and feldspar (aluminium potassium silicate)	Easily fractured, can withstand high temperatures, good resistance to chemical corrosion	Bricks, tiles, sanitary ware, roof tiles, space shuttle insulation tiles, engine blocks
Glass	Silica, sand, lime and sodium carbonate	Brittle but can be tempered, varieties can be coloured with oxides, lead glass has high refractive properties, borosilicate glass or 'pyrex' can withstand high temperatures	Cookware, optical lenses and magnifiers, windows, mirrors
Concrete	Cement, sand, **aggregate** (small stones) and water	Very high compressive strength	Structural beams/members, road surfaces, pathways and drives, foundations

EXAMINATION QUESTION

Example question and answer.

 Q1 *Describe how concrete is made and strengthened, and give an industrial application of its use.* **(5 marks)**

Acceptable answer
Concrete is made from a combination of cement, water and aggregates. Although the strength of the concrete can be **controlled to a certain extent by varying the ratio of cement to aggregate,** its strength can be greatly enhanced by the **addition of steel reinforcing bars.** This is

especially so when **concrete beams are used as lintels above door frames.** The **reinforcing bars are placed towards the bottom of the section and run along the entire length of the lintel.**

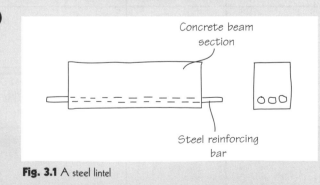

Fig. 3.1 A steel lintel

Classification of components

KEY TERMS

Check you
understand these terms

threads, permanent joints

Further information can be found in *Advanced Design and Technology for Edexcel, Product Design: Resistant Materials Technology*, **Unit 3A, section 1.**

KEY POINTS

Nuts and bolts

A nut is a collar, usually made from steel, that has a threaded hole through the middle that fits onto a threaded bar or bolt. Bolts are made with a hexagonal head at one end with a screw thread on the other.

Although many forms of screw **threads** exist, they have three main uses:

• converting rotary to linear motion
• obtaining a mechanical advantage
• fastenings.

The most common form of screw thread is the V-thread. This can be used and cut both internally and externally and is a relatively easy process to carry out in the school workshop. To cut an internal thread, a hole is drilled before a tap, which is held in a tap wrench, is wound into the hole. External threads are cut onto bars or rods with a die held in a die stock. Great care must be taken to ensure that the tap or die are perpendicular to the bar, rod or hole being tapped. Care should also be taken not to use too much force since the taps and dies are made from high carbon steel and are likely to break if forced. A cutting lubricant can be used to aid with the tapping and threading procedures.

Rivets

Rivets are used to make simple **permanent joints** between two or more pieces of metal.

Table 3.9 Screw thread forms

Thread type	Thread form	Applications
V-thread		General purpose, fastenings, nuts and bolts
Square		Allows large forces to be applied and is used in vices and cramps
Buttress		Used where force only needs to be applied in one direction, such as woodwork vices. Used in association with quick-release mechanisms
Acme		Used on a centre lathe where the engaging nut must clamp onto a rotating shaft. This allows the tool post to move automatically along the lathe bed

They can form rigid joints and hinged joints, for example, in a pair of pincers.

Snap rivets are normally made out of soft iron, which is ductile and can be easily deformed with a hammer to form the joint. They are available with three different head types: countersunk, flat, and snap or round headed.

Pop rivets are made from aluminium with a central steel pin. A special pop rivet gun is used to de-form the rivet. They are especially suited to joining thin sheet material or where access can only be gained from one side of the joint.

Gears

A gear is a toothed wheel with a special profile that allows it to mesh with other gears. This allows forces to be transmitted and the rotational direction changed. When two gears mesh, they form a simple gear train; the rotational direction of the output will be opposite to that of the input. When the gears are different sizes, the rotational speeds will also differ. An idler gear can be used with a simple gear train to ensure that both the input and output directions are the same. As more gears are introduced into the train, a compound gear train is formed, giving larger increases or reductions of speed and rotational force.

Bushes and bearings

Any rotating shaft must be supported at least at one end. Bearings and bushes provide a means of support that allows the shaft to rotate freely. A bush is the simplest form of bearing; it is essentially a cylindrical sleeve that fits into a hole through which the shaft fits. Bearings exist in many forms, the commonest types being plain, roller and ball bearings. Bearings are made from brass, bronze, white metal or some types of plastic, e.g. nylon.

Cams

Cams are mechanisms normally used for converting rotary motion into reciprocating and oscillating motion. The cam is fixed to a rotating or oscillating shaft; a follower is held against the cam and moves as the cam rotates. The main types of cams are pear-shaped, eccentric and heart-shaped. The main types of followers are knife-edge, flat, roller and point.

Stock sizes

Stock sizes are the sections and sizes in which materials are produced. When designing products, it is important to consider what sizes are available so that the most efficient use can be made of material. For example, MDF is available in sheets 2440 × 1220 mm, so careful planning when making a desk will ensure that little waste is left.

Table 3.10 Gear forms and applications

Gear form	Application
Gear train	Reduction/increase of speed – change of direction in children's toy cars
Worm and wheel	Changing rotary motion through 90 degrees – food mixers, electrical drills; large speed changes achieved
Bevel gears	Changing rotary motion through 90 degrees – workshop hand drill
Rack and pinion	Converts rotary into linear motion – lock gates, pillar drill

EXAMINATION QUESTION

Example question and answer.

 Q1 *Screw threads exist in many forms. Name **one** type of screw thread and give an application where it is used.* **(2 marks)**

Acceptable answer
A **square thread** is one type of thread form. It is used in **sash and G-cramps** where large forces are needed to hold things while they are being glued and dried.

Working properties of materials and components

You need to

know about working properties and functions of materials and components relating to the composition and structure of materials:

☐ alloying metals to change properties
☐ mechanical properties
☐ heat treatment
☐ work hardening.

KEY TERMS

Check you
understand these terms

alloy, mechanical property, heat treatment, work hardening

Further information can be found in *Advanced Design and Technology for Edexcel, Product Design: Resistant Materials Technology*, Unit 3A, section 2.

KEY POINTS

Alloying metals to change properties

An **alloy** is produced by combining a metal with one or more other elements. Creating an alloy gives the new material properties that none of the original metals possessed. Alloying can:

- change the melting point
- increase mechanical properties, such as hardness

- change physical properties, such as electrical conductivity
- enhance resistance to corrosion
- change colour
- improve fluidity in order to create better castings.

Mechanical properties

A **mechanical property** is associated with how a material reacts when a force is applied to it. The material will either behave elastically, returning to its original shape and size, or it will behave plastically, deforming permanently. If the material is taken beyond its plastic limit, deformation ultimately ends in the material or component breaking.

Heat treatment

Heat treatment is the process of heating and cooling materials in a controlled fashion in order to change their properties and characteristics (Table 3.13 on page 43).

Work hardening

As a piece of metal is deformed by cold working, such as hammering, bending or rolling, **work hardening** occurs and deforms the grain structure.

Age hardening is mostly confined to aluminium alloys. Although some heating is involved, hardening increases afterwards, when the material has been cooled or 'aged'.

Table 3.11 Alloys and their applications

Alloy	Composition	Properties/applications
Mild steel	Fe + 0.15–0.3% C	Tough, ductile and malleable. Cannot be hardened and tempered. Nails, nuts and bolts
High carbon steel	Fe + 0.8–1.5% C	Very hard and less ductile. Easily heat treated. Hammers, chisels and screwdriver blades
Stainless steel	Medium carbon steel + 12% Cr + 8% Ni	Hard and tough. Corrosion resistant. Cutlery and kitchen sinks
Brass	65% Cu + 35% Zn	Casts well. Good corrosion resistance. Boat/plumbing fittings

Table 3.12 Material properties

Property	Description	Material	Applications
Strength	Is broken down into five types: – compressive – tensile – shear – torsional – bending		
Elasticity	The ability to return to its original shape once the deforming force has been removed	Rubber, ash, spring steel	Springs, tennis racquets
Plasticity	Ability to be changed permanently by heating, without cracking or breaking	Acrylic, ABS	Baths, moulded products, car indicators
Ductility	Ability to be drawn or stretched	Silver, copper, aluminium	Electrical cables
Hardness	Ability to withstand indentation and abrasion	High carbon steel	Drill bits, taps and dies
Malleability	Ability to be deformed by compression without tearing or cracking	Lead, low carbon steel	Roof flashing, car body panels

Table 3.13 Heat treatments

Process	Description
Hardening in steels containing more than 0.4% C	The metal is heated above its upper critical temperature, which is 720°C for a steel containing 0.83% carbon. It is held or 'soaked' at this temperature in order to achieve a uniform heat throughout. It is then quenched in water, oil or brine
Tempering	Once a metal has been hardened, it also becomes very brittle. In order to remove the brittleness, the material has to be tempered. The metal is cleaned and the surface polished before the metal is heated. The temperature is dependent upon the application. For example, a screwdriver blade would be heated to 290°C – a blue colour
Annealing	This process is used to restore the crystal structure of a metal after it has been work hardened. For example, as copper pipes are drawn through smaller and smaller dies, the metal becomes harder. The metal is heated and soaked and then allowed to cool very slowly in air
Normalising	In forged components where the grain size may be irregular after forging, the component is normalised. This involves heating the work piece to just above its upper critical temperature before allowing it to cool in still air. This ensures a more refined grain structure, and the piece will now be more ductile and possess greater toughness

Hand and commercial methods

You need to

know about hand and commercial methods of preparing, processing, manipulating and combining materials and components to enhance their properties, including associated tools, machinery and equipment, including CAD/CAM in relation to:

- □ **fabrication/manufacture**
- □ **processes**
- □ **removal**
- □ **addition**
- □ **joining.**

KEY TERMS

Check you understand this term

fabrication

📖 **Further information can be found in *Advanced Design and Technology for Edexcel, Product Design: Resistant Materials Technology*, Unit 3A, section 3.**

KEY POINTS

Fabrication/manufacture

The term **fabrication** is used to describe the process of joining materials together. Table 3.14 indicates the types of fabrication processes that can be applied to various materials.

Processes and removal methods

Casting

Casting involves pouring molten metal into an empty cavity. Casting can be carried out in two ways: die casting and sand casting.

Die casting uses a metal mould made from several parts, enabling the mould to be used repeatedly. Die casting is normally used for aluminium and zinc alloys, which have relatively low melting points. Components made by die casting have a high quality surface finish.

In sand casting, a former or pattern of the required shape is made first. Sand is packed very tightly around the pattern, which is removed to leave a cavity. Runners and risers

Table 3.14 Fabrication techniques

Wood	Metal	Plastic
Joints: frame/box/edge	Welding	Ultrasonic welding
Laminating	Soldering	Fusion welding
Nails/screws	Rivets	
Knock-down fittings	Nuts and bolts	

allow the molten metal into the mould and the air out. Once the metal has solidified, the cope and drag (the metal container into which the sand was packed) is broken open and the cast product removed. In most cases, the sand is cleaned and reused. The runners and risers are cut off and recycled and the formed product prepared.

Sintering

Sintering is used to form components from metals and other elements with very high melting points and from metals that are not miscible in the liquid state. The powders are mixed before being compacted in a shaped die under very high pressure. A certain amount of heat is applied to help fuse the particles together. Sintered components are often used for bearings and bushes.

Shearing

Shearing is a form of cutting. Hardened steel blades with ground cutting edges are brought to bear on the material and as more pressure is applied the material is cut. Scissors, tin snips and guillotines all use the shearing action to cut materials.

Stamping

Stamping is a process in which plastic deformation takes place by compression. A die is used, which normally comes in two parts. A certain amount of work hardening takes place as the components are stamped out and in some cases heat treatment has to be carried out to restore the material's original grain structure. Keys, coins, medals and radiator panels are all produced by stamping.

Milling

Milling is a wasting process, which can be used on metals and plastics. The piece of work is clamped to a table and moves past a multitoothed rotating cutter. Milling can be carried out vertically or horizontally, but regardless of which method is being used, the table can move in three directions or axes: X, along the length of the table, Y, across the width, and Z, up and down vertically.

With the development of CNC (Computer Numeric Control), milling machines can now be used repeatedly to cut identical components for high volume production or they can be used to produce very complicated one-off components that would have been impossible to do manually.

Drilling

Drilling is used to cut holes with a rotary cutting tool. The pillar drilling machine is the most common machine found in school workshops and can take a variety of cutting tools. A twist drill, the most frequently used drill, is made from high-speed steel (HSS).

Turning

Both metal and wood can be turned easily. In each case, the piece of material is rotated and the tool is held against it and moved. When turning metal on a centre lathe, the tool is held in a tool post and is moved by controlling a series of wheels. Wood turning requires the user to hold the tool, although it is supported on a tool rest. A number of processes are identical for woods and metals, such as facing, a cutting process perpendicular to the work, parallel turning, cutting parallel to the work, and taper turning, providing a taper on the work. Different tools and shaped tools are used for different processes, such as roughing out, finishing, parting off and shaping. As with milling, CNC machines are now widely used in both schools and industry. These allow the speedy production of identical and complicated shapes, although the set-up costs are high.

Grinding

The grinding of metals can be carried out by hand with a disc, offhand or surface grinder. All methods use a disc made from an abrasive powder cemented together. Grinding is used to clean up welds or to prepare the surfaces of cast products and components.

Spark erosion

Spark erosion is used to machine very hard materials and complicated shapes and profiles. A small spark, generated at up to 10,000 times a second, jumps between two conductors, one being the tool and one the piece of material being eroded. As each spark is produced, a tiny piece of metal is eroded away. Spark erosion is used to produce dies for extrusion, injection moulding and blow moulding.

Laser cutting

Lasers are used to cut metals and other materials, such as textiles. This is a very accurate method of cutting and very complicated shapes and profiles can be achieved. When cutting metal, a jet of oxygen is used to blow away the molten metal and the oxide film that forms on the surface.

Forging

Forging involves heating the metal piece and hammering it into shape. It is sometimes quicker to forge components than to machine them from solid metal. A drop forge uses a mould in two halves. The hot metal piece is placed into the bottom half of the open mould and the top half is dropped with an enormous force. Drop forging is used to make hip replacement joints.

Injection moulding

Injection moulding is a very highly automated process used to produce a wide range of plastic items, such as product casings, washing-up bowls and buckets. Although the process is quite simple, the products are manufactured to a very high quality and generally require no surface finishing.

Addition and joining

Joining methods can be categorised into three main groups.

- Permanent – once made they cannot be reversed without causing damage to the workpiece.
- Temporary – although not always designed to be taken apart, they can be disassembled if needed without causing damage.
- Adhesives – most adhesive bonding can be classified as a chemical reaction (see Table 3.15).

Table 3.15 Adhesives

PVA	A white woodworking glue, which is easy to use and provides a strong joint, providing the joint is a good fit. Not all types of PVA are waterproof
Cascamite	Stronger than PVA but it is supplied as a powder that needs to be mixed with water. It is, however, waterproof and is widely used for outside furniture and boatbuilding
Epoxy resin	Comes as a two-part package – a resin and a hardener. The two parts have to be mixed in equal amounts and it takes a little while to set. It is expensive
Tensol cement	Available in various forms, but Tensol 12 is the most common. Used for sticking/bonding acrylics

Rivets

Rivets are used extensively in sheet metal work and can also be used to join acrylic and wood. Soft iron rivets are available with different head shapes. Pop rivets are hollow and used extensively in the aircraft industry.

Screws

Screws are regarded as a temporary method of joining unless they are used in association with an adhesive. Screws are classified by length, diameter and head type, and they are available in a variety of materials and finishes.

Nuts and bolts

Nuts and bolts are a method of temporary fixing. Bolts are mostly hexagonal headed and are used with matching thread forms to form very strong mechanical joints.

Heat processes

All heat processes used for joining are regarded as permanent fixing methods (Table 3.16).

Table 3.16 Heat processes

Process	Description
Soldering	Soldering is generally only used in light fabrication work but it can be broken down into different types: Soft – 200°C: used for joining brass, copper and tin plate. Hard – 625°C: gas torches are needed to generate the heat needed but joints are much stronger. Silver – 625–800°C: allows joints to be made in stages since subsequent joints can be made at lower temperatures.
Brazing	This is a type of soldering and is sometimes referred to as hard soldering. A filler rod, brazing spelter, is used to make the joint. Fluxes must be used to aid the flow of braze.
Welding	Welding is carried out at such high temperatures that the filler rod, the parent metal, melts into the joint to make an exceptionally strong joint. Welding exists in many forms: oxy-acetylene, arc, spot, MIG (metal inert gas) and ultrasonic. Ultrasonic welding can be used on plastics as well as metals. A probe is placed in contact with the surface of the material; ultrasonic pulses passing through the probe cause the temperature to rise and as pressure is applied a bond is formed.

EXAMINATION QUESTION

Example question and answer.

Q1 *Materials can be joined by heating processes, such as:*

- *soldering*
- *brazing*
- *welding.*

Briefly describe **one** *of the above processes.* **(4 marks)**

Acceptable answer

Welding is a process that can be carried out in a number of ways, all **involving very high temperatures**. In most instances, **a filler rod of the parent material is melted and used to join the two pieces**. As a result of the temperatures involved, **the area being joined melts along with the filler rod. On cooling, the welded area solidifies** and makes for a very strong joint.

Finishing processes

You need to

know about applied finishes to improve quality and provide enhanced aesthetic or functional properties including:

- ☐ surface coating
- ☐ self-finishing
- ☐ surface decoration
- ☐ the relationship between finishes, properties and quality.

 Further information can be found in *Advanced Design and Technology for Edexcel, Product Design: Resistant Materials Technology,* **Unit 3A, section 3.**

 KEY TERMS

Check you understand these terms

anodising, painted, varnish, self-finishing

KEY POINTS

Surface coating

Any finish is used extensively to improve the product's functional properties, aesthetic qualities and generally serve to improve the overall quality. This relationship must be considered early on by the designer as one of the major design considerations.

Anodising

Anodising is essentially used for aluminium and its alloys. The whole product is immersed in a sulphuric acid solution. The product being anodised is made an electrical anode and lead plates in the tank are used as cathodes. When a direct current (DC) is passed through the solution, a thin oxide film forms on the product surface. Coloured dyes can be added before the surface is finally lacquered.

Painting

A **painted** surface finish can exist in many forms. However, careful and thorough surface preparation is essential. Metals must be degreased before being primed, undercoated and finally topcoated. Woods, especially knots, should be sealed in order to prevent resin escaping. It is essential that between each coat, whether on wood or metal, the surface is lightly rubbed down. Commercially, much of the painting is done by spray painting, but great care must be taken to ensure an even and consistent application over the entire surface. Topcoats are available in three main types (Table 3.17).

Table 3.17 Paint types

Paint type	Description and applications
Oil-based	Commonly known as gloss, general interior use for doors, skirting boards, door frames
Emulsion	Available in vinyl or acrylic but they tend not to be waterproof; walls and ceilings
Polyurethane	Tough and scratch resistant, widely used on children's toys

Varnishing

A **varnish** is a plastic type of finish made from synthetic resins. Varnishes offer a tough waterproof finish in either a gloss, matt or satin finish. New varnishes are often acrylic-based and dry more quickly and smell less.

Self-finishing

Plastics generally require little or no finishing on the edges or surfaces and are readily described as being **self-finishing** due to the nature of the material and processes involved (such as blow moulding or injection moulding). Textures and surface patterns can be introduced to the mould surface and will subsequently replicate in the surface of the finished moulded product.

Surface decoration

Most materials can be engraved, although engraving is usually on glass, metal goblets, plates and rings. Special high-speed tools and bits are used on glass whereas very fine chisels and scrapers can be used on metals. Glass can also be sandblasted to achieve some very deep surface penetration.

Vinyl film is now widely used by sign writers to create large and colourful signs and vehicle graphics. The data is prepared on CAD packages and is output on large plotter cutters before being transferred to the product.

The relationship between finishes, properties and quality

Whatever type of finish is used, its main purpose is to improve the product's functional

Fig. 3.2 A plotter cutter

properties, aesthetic qualities and generally serve to improve overall quality. A hanging basket bracket made from mild steel would quickly rust and tarnish when left outside if it was not finished. Plastic dip coating would be used to protect the surface and to enhance the overall quality of the product.

EXAMINATION QUESTIONS

Try answering these yourself.

 Q1 *Give **two** reasons for applying a surface finish, with examples.* **(4 marks)**

 Q2 *Describe **two** advantages of manufacturing vehicle graphics on a CAD system.* **(4 marks)**

Product manufacture

You need to

- [ ] **know about one-off, batch and high volume manufacture**
- [ ] **know about the impact of different levels of production on costs and production time.**

 KEY TERMS
Check you understand these terms

one-off, batch, high volume

Further information can be found in *Advanced Design and Technology for Edexcel, Product Design: Resistant Materials Technology,* **Unit 3A, section 3.**

KEY POINTS

One-off, batch and high volume manufacture and the impact of different levels of production on costs and production time

The scale of production is a key factor that any product designer must consider. The eventual level of production will fall into one of three categories.

One-off production

A kitchen to fit a specific space and the new Wembley Stadium are both examples of one-off design and manufacture. They are both unique solutions and costs are high because specific components have to be made. Production time will be prolonged due to the intensive and specialised labour required.

Batch production

Batch production is the production of identical products of specified quantities in 'batches', varying from ten to several thousands. The necessary tooling is expensive to set up but is kept as flexible as possible. Once a batch has been completed, the machine is used for another product, but can be re-configured at a later stage if needed. Kitchen doors and drawer fronts are produced in batches, despite being finished in a number of ways and colours. Production times are lower for batches since once a machine has been set up or programmed it can be kept running, reducing the unit cost.

High volume production

High volume, or mass, production often runs continuously for 24 hours a day. High investment in machines and tools is required but tens of thousands of identical components are produced, having a significant impact on the unit cost. Minimal skilled labour is required since the machines are highly automated, needing only a very few skilled operators to set up and maintain them.

EXAMINATION QUESTION

Example question and answer.

Q1 *Give **one** example of a batch produced product. Explain why this scale of production is appropriate.* **(5 marks)**

Acceptable answer

An **Olympic gold medal** would be considered as a batch produced item. In any one Olympic Games, it is only likely that some **150 events take place and subsequently only this number of gold medals would be required**. It is likely that they are **cast or pressed/stamped and as such a die or mould would be required**. Once this is made by a **highly skilled worker, a less skilled operative would be capable of making the medals**.

Testing materials

You need to

☐ **understand the purpose of British, European and international standards**
☐ **understand the need for safety testing under controlled conditions**
☐ **understand the use of ICT in testing**
☐ **understand ultrasound testing.**

KEY TERMS

Check you understand these terms

virtual reality, ultrasound

Further information can be found in *Advanced Design and Technology for Edexcel, Product Design: Resistant Materials Technology*, Unit 3A, section 4.

KEY POINTS

The purpose of British, European and international standards

Products must be safe to use before they can be sold. The testing of products is strictly governed by the British Standards Institute (BSI) and other International Standards Organisations (ISO) around the world. The British Standards Institute is one of the largest independent testing organisations in the world and is responsible for the testing of products from medicine bottles to children's toys. Products that pass the tests are awarded a safety mark.

Safety testing under controlled conditions

All testing must be carried out under controlled conditions otherwise any subsequent testing will not be consistent and fair. For example, all chairs and tables should be subjected to the same load and tipping tests.

The use of ICT in testing

More and more testing is now carried out using information and communications technology (ICT). As computer modelling has become more advanced and powerful, programs are used to simulate what would happen to products if hit, dropped or squashed. This has led to financial savings since fewer 'live' products and components are now tested to destruction.

Virtual reality has also become increasingly popular. This allows architects to show how their buildings would look, both outside and inside.

ICT also allows simulations of product performance to take place prior to production. This allows colour and texture to be viewed, as well as looking at the simulation of manufacturing times, tool paths and clamping points.

Ultrasound testing

Ultrasound testing involves passing a frequency of between 500kHz and 20MHz through a component. This process generates a picture that identifies any flaws inside the component. Ultrasound testing is used extensively to check welds and cast products.

EXAMINATION QUESTION

Example question and answer.

 Explain, giving an example, why testing must be carried out under controlled conditions. **(2 marks)**

Acceptable answer

Any testing must be carried out under controlled conditions to ensure that **tests are consistent and fair**. For example, **a child's cycle helmet would probably undergo a range of tests, one of which would be a form of impact testing.** It would be **unreasonable to test two different cycle helmets by exposing them to two different impact loads since this would naturally lead to one performing better than the other and passing, giving an unfair advantage to the manufacturer.**

PRACTICE EXAMINATION STYLE QUESTION

1   Select and describe, using notes and sketches, **one** of the following:
    • blow moulding
    • injection moulding
    • vacuum forming. **(4)**

2   Surface decoration can be applied in several ways:
    • engraving
    • etching
    • vinyl transfer
    • pyrography.
    Select **two** of the above methods and describe how they are used to decorate the surface
    of a product, giving an example of each. **(6)**

3   Properties of metals can be changed or modified by:
    • alloying
    • hardening
    • tempering
    • annealing.
    Briefly describe **two** of the above processes. **(4)**

4   Name **three** different types of cams, giving an application of each. **(6)**

5   Gears are available in a range of forms:
    • worm
    • crown
    • bevel
    • rack and pinion.
    Select **one** of the above and briefly describe, using notes and sketches, how they work
    and give an application. **(4)**

6   Metals can be divided into two main groups:
    • ferrous
    • non-ferrous.
    Give an example of a material from each group and a different working characteristic
    from each group. **(4)**

3 B1 Design and technology in society (R302)

This option will be assessed in Section B during the 1½ hour, Unit 3 examination. If you have chosen this option, you should spend half of your time (45 minutes) answering all of the questions for this section. It is important to use appropriate specialist and technical language in the exam, along with accurate spelling, punctuation and grammar. Where appropriate, you should also use clear, annotated sketches to explain your answer. *You do not have to study this chapter if you are taking the CAD/CAM option or the Mechanisms, Energy and Electronics option.*

The physical and social consequences of design and technology in society

You need to

understand the effects of design and technological changes on society:

☐ mass production and the consumer society
☐ the 'new' industrial age of high-technology production
☐ the global market place
☐ issues related to global/local production.

understand the influences on the development of products:
☐ product reliability and aesthetics
☐ design and culture
☐ new materials, processes and technology.

KEY TERMS

Check you understand these terms

consumer society, global market place, global manufacturing, Arts and Crafts, Memphis, Art Nouveau, Bauhaus, Art Deco, The New Design, new materials, smart materials, eco-design

Further information can be found in *Advanced Design and Technology for Edexcel, Product Design: Resistant Materials Technology*, Unit 3B1, section 1.

KEY POINTS

Mass production and the consumer society

The Industrial Revolution and mass production changed the way in which design and technology influenced society. A gradual, evolutionary pattern of development in technology and society was replaced by rapid technological and social change, which has led to positive and negative consequences (Table 3.18).

Mass production

The history of design began with the Industrial Revolution and the invention of the steam engine in the mid-1700s. Coal mining, iron and steel and machine production took on a new importance and set the scene for the development of industrial mass production. Until the Industrial Revolution and the advent

Table 3.18 Some advantages and disadvantages of technological progress

Some advantages of modern technology	Some disadvantages of modern technology
More consumer choice	Environmental damage due to pollution
More affordable products	Social problems caused by globalisation
Higher living standards	Influence of computer games on children

of powered, automated machinery, production was limited to the quantities of products that could be produced by craftsmen by hand. Until the invention of modern printing techniques, for example, texts had to be laboriously copied by hand.

Mass production simplified the production process by dividing it into simple, repetitive tasks.

- Many traditional trades could not compete, leading to unemployment.
- Unskilled jobs led to lower wages.
- Poverty led to social tensions, strikes and uprisings.
- The movement of labour from the countryside to the cities led to many social problems.
- Unregulated development led to environmental problems, such as pollution.

The concerns of the workers led to the birth of the modern trade union, which fought to combat these problems.

Assembly lines

Assembly lines transported the work to the workers within the factory and vastly accelerated the production process. Products became cheaper and more widely available to more sections of society.

Design tradition and design for mass production

Early, mass produced products tended to imitate traditionally produced styles. It was not until the second half of the nineteenth century that designers began to reflect the new social conditions and new production processes. The heavy, over-ornate styles were replaced by products that reflected the needs of the market. Designers began to recognise the market potential of the new class of people created by the Industrial Revolution. Simple, inexpensive consumer goods began to be produced for the expanding working classes.

Design in the USA

The lack of competition and drive to mass produce products meant that designers concentrated upon the functional and technical aspects of products, often at the expense of styling.

The development of the consumer society

It was not until the twentieth century that fashion and styling became important as a

Table 3.19 The effects of some political, social and technological developments on consumer society

Developments	Effects
World conflicts	Fuelled demand for new products and new technologies
Introduction of the national grid (1920s)	Stimulated demand for new, 'labour-saving' electrical appliances
Increasing standard of living	Higher demand for luxury goods and travel
Growth in media industry	Growth of advertising, packaging and design industries, which fuelled consumer demand
Development of new materials and technologies for space exploration	Allowed new products to be produced, e.g. streamlining in the automotive industry, which reflected advances in aerodynamics research and the desire to appear modern

dominant feature of product design. A rise in incomes and change in attitudes to consumption led to the growth of the **consumer society** as society demanded more products. People no longer bought products just for their function.

Designers began to develop new features in their products.

- Packaging became a much more important part of the product.
- Products began to be designed with limited lifespans (planned obsolescence) to be replaced with restyled or improved variants.
- Styling became significantly more important as competition between products increased.

The 'new' industrial age of high-technology production

The latter stages of the twentieth century have seen an ever more rapid pace in technological development with the introduction of new materials and new processes.

- Computer technology: the development of the silicon chip in the 1960s had a profound effect on society.
- New materials, such as aluminium, stainless steel, heat resistant glass and modern polymers, have made new products possible.
- New technology, such as miniaturisation, which allows the functions of different devices to be combined into a single product, has led to a consumer trend of being attracted to the latest technological products.
- Fashion: as new technologies become increasingly affordable and accessible, styling becomes a more important factor in attracting consumers.

Examiner's Tip

Familiarise yourself with a range of high-tech products so that you can talk about them in the exam. You need to be able to discuss the development of these products by referring to the older products that they have replaced.

Fig. 3.3 A modern Swatch watch

Case study: the Swatch watch

Originally developed in the 1980s, Swatch watches were targeted at the lower end of the market. The company produced affordable watches that were heavily styled and aimed to reflect contemporary fashions. Swatch watches today have switched their emphasis by concentrating on integrating new technology. Watches have been produced that can be used as electronic ski-passes or travel tickets. More recently, the company has started to develop models that can be used to access Internet services or act as a mobile phone.

The global market place

In order to remain competitive, companies who operate in the **global market place** (operate in different countries) need to ensure that their products appeal to a wide range of people from different cultures. Some companies employ designers in different countries, while others are careful to conduct thorough market research in unfamiliar niche markets. Products may be remodelled to reflect differences in fashion, taste, legal requirements and other cultural differences. Remodelling may involve:

- changing a product name to take account of linguistic differences
- increasing the amount of recyclable materials used in the product
- a decision as to whether to fit a plug or not
- the fitting of different visual displays
- using devices to suppress noise levels.

Global manufacturing

Global manufacturing is associated with multinational companies who rely on overseas manufacturing capacity.

- Global manufacturing has only become possible through developments in international communications technologies and is attractive to multinational companies because it allows them to take advantage of much lower labour and materials costs.
- Products can be designed and developed in one country, manufactured in another and shipped all over the world.
- The flexibility of modern industry means that it is easy to switch production between countries.

The trend towards global manufacturing often involves designing products, such as books or computer games covers, in one country and printing them in another. Design studios, equipped with state of the art computer technology and broadband, can receive briefs in any format, speeding up the turnaround from concept to the finished design. Systems are online 24 hours a day to allow continuous contact between clients, the design team and the film planning, plate making, printing and finishing departments.

Examiner's Tip

You should familiarise yourself with a range of global manufacturers so that you can use them to illustrate your answers in the exam.

Issues related to global/local production

Table 3.20 The advantages and disadvantages of global manufacturing

Advantages of global manufacturing for NICs* and LEDCs**	Disadvantages of global manufacturing for NICs and LEDCs
• Higher employment levels and living standards. • Improved the level of expertise of the local workforce. • Source of foreign currency improving the balance of payments. • Widening of the economic base. • Transfer of technology possible.	• Possible environmental damage to unspoilt areas. • Jobs require only low-level skills. • Managerial posts are filled by people from MEDCs.*** • Profits are exported back to MEDCs. • Poor legislation or enforcement allows multinationals to cut corners on health, safety and pollution. • Multinationals can exert political pressure. • Raw materials are often exported with no value added. • Manufactured goods are exported and do not benefit local community. • Decisions are made in a foreign country and on a global basis. • Reduced need for the local workforce due to automation.

*NICs: newly industrialised countries, such as Singapore and Taiwan
**LEDCs: less economically developed countries, such as many nations in Africa or Asia
***MEDCs: more economically developed countries, such as most of Western Europe and North America

Influences on the development of products

Important changes have emerged as a result of improving standards of living and changes to industrial organisation, including the emergence of the 'professional designer' and the growth in importance of fashion and style. The modern designer needs to consider that:

- form has become as important as function for many products

1906 – Strowger Calling Dial, Strowger USA

1931 – Siemens telephone, designed by Jean Heiberg

1970s – Standard GPO British telephone

2001 – Nokia 9210; a modern mobile phone able to access the Internet and email

Fig 3.4 The development of the telephone

Fig 3.5 Philippe Starck's lemon squeezer

Fig 3.6 A traditional lemon squeezer

- designs must be 'market led' or 'market driven', reflecting the needs and tastes of the consumer
- customers buy products that reflect their aspirations
- products need 'personality' to make them stand out from competing products.

Product reliability and aesthetics

Product reliability is no longer an issue as most products carry guarantees, but designers consider the aesthetic qualities of a design carefully. If an otherwise well-designed, high-performance product is painted the 'wrong' colour, for example, few people will buy it. Designers need to think carefully before making aesthetic decisions about shape and form, line, balance, colour, decoration, surface pattern, scale, styling, and texture.

Form and function

Form and function can place conflicting pressures upon the development of a design. The most successful designs both 'do the job well' and 'look good'. The history of design has seen many debates over the relative importance of form in relation to function, for example:

- the **Arts and Crafts** movement of the nineteenth century reacted against the highly decorated forms of mass produced Victorian products
- postmodernist movements, such as the **Memphis** group, reacted against the pure functionalism of Modernism.

Modern products such as packaging can be said to have three basic functions.

- The practical and technical function: does the packaging contain, protect and preserve the product?
- The aesthetic function: does the packaging have visual appeal to attract the consumer?
- The symbolic function or the image it gives the user: does the packaging project a desirable brand identity?

Products usually evolve gradually because it is easier to avoid mistakes and control development costs.

Examiner's Tip

Remember that you will need to revise the whole specification. It is quite possible for a topic to be repeated for a second year and it is quite possible that a topic could be omitted for a couple of years.

Design and culture

The designer needs to work within the constraints of:

- the design brief (or specification)

- available materials
- production technologies
- the expectations of the consumer in terms of appearance or style.

The most successful designers are able to establish 'new' styles, which are often adopted and transferred to different products. The Dyson vacuum cleaner, for example, adapted existing cyclone technology to domestic use. The styling of these new cleaners was boldly different, using bright colours and transparent plastics.

Examiner's Tip

You need to learn about the history of important design movements from different periods. It is unlikely that the outlines below will provide you with enough material to answer the questions fully. Read around the subject in preparation for the examination.

Arts and Crafts

The Arts and Crafts movement was founded in 1890 by the English artist, designer and writer William Morris.

- The movement was founded as a reaction against the effects of industrialisation and mass production of the Victorian era.
- The concepts of art, craftsmanship and quality were emphasised during production, which used natural materials and traditional techniques.
- Decoration was uncomplicated and based upon simple, organic forms from nature.
- The movement was involved in all areas of design from textiles to typefaces.
- Many modern European designers have been influenced by the Arts and Crafts movement.

Art Nouveau

Art Nouveau was an important international movement that developed during the late nineteenth century. Charles Rennie Mackintosh was inspired by Japanese geometric forms.

- Designs are characterised by the use of organic, free-flowing lines and shapes.

- The movement made use of traditional craftsmanship and mass production.

Modernism and the Bauhaus

Walter Gropius founded the **Bauhaus** school early in the twentieth century (1919 and 1933). Laslo Maholoy-Nagy created simple geometric typestyles. Famous artists, such as Paul Klee, taught at the school. The Nazis closed the school in 1933, which encouraged the spread of Bauhaus ideology around the world.

- Function determined form in many Bauhaus products.
- Artistic education and crafts training were considered equally important.
- Modern materials and processes were widely used.
- Products were designed to be suitable for mass production.
- Bauhaus was, and is, very influential, inspiring the 'International Style' in architecture.

Art Deco

From 1925 onwards, the **Art Deco** style became extremely popular and influential. It was named after an international exhibition in Paris in 1925 in which all exhibits were required to be novel in their design. It was influenced by modern paintings and a popular interest in African and Egyptian art.

- The style is characterised by many visual elements, such as bright colours, images of the sun, geometric shapes and zigzag patterns derived from Egypt.
- There was an emphasis on the use of expensive materials, such as ebony, ivory and bronze.
- It influenced the design of many products and made use of new materials including aluminium, plywood and Bakelite (an early thermoset plastic).

The 1940s and 1950s

In the 1950s, the development of the supermarket meant that food packaging became a marketing tool. Packaging had to be instantly recognisable to the consumer. 'Italian style' products became all the rage. The 1950s saw the development of new plastics, such as foam, nylon and polyester, arising out of wartime research.

- New materials, technologies and processes inspired new, mass produced, low-cost products.
- Images of new developments in science and technology, such as the splitting of the atom, were popular.
- New technologies inspired new graphic forms, such as television graphics, information graphics and corporate identity systems.

Youth culture

The 1960s and 1970s were a period of rapid political, social and economic change.

- Young people became a new and important market.
- Youth culture inspired new styles, such as 1960s 'psychedelic' art.
- The growth of the media led to the development of marketing and advertising.
- New scientific advances, such as those driven by the space programme, introduced new materials.
- New futuristic products were inspired by science fiction television programmes and films.
- Improved production techniques enabled designers to tailor products to small niche markets.
- Pop music and the 'hippie' movement influenced graphic design, fashion and interior design.
- Artists, such as Andy Warhol ,were influenced by graphic styles and in turn influenced new graphic designers.
- Packaging became bolder, using vivid colours and strong visual images.
- New materials, such as cellophane, aluminium and plastics, enabled the development of new types of packaging, such as disposable, ring-pull cans, Tetrapak™ cartons and moulded plastic containers, which were lighter and cheaper to transport than fragile glass.

Memphis

Memphis was the name of a group of designers who established themselves in Milan in Italy in 1981. Ettore Sottsass was an architect who moved into product design and became the principal figure of the movement.

- The Memphis designers were interested in mass production, advertising and the practical objects of daily life.
- Designs were colourful, witty and stylistic, influenced by comic strips, films and punk music.
- Products were designed with unusual combinations of materials, such as melamine, glass, steel, industrial sheet metal and aluminium. Many products looked like children's toys.
- The 1980s saw the status of design grow. Design took over a key role in the development of individual lifestyles.

Design after Memphis

The New Design of the recent era was led by designers such as Ron Arad, Jasper Conran and Tom Dixon.

- Many designers moved away from functionalism and aimed to reflect the influences of daily life.
- Designers in Germany and the UK simplified forms and began working with materials such as concrete and steel.
- Designers made use of unusual combinations of colour and modern materials.
- Designers became more environmentally aware, designing products using recycled or recyclable materials.
- Marketing in the 1980s and 1990s developed into a much more significant force within product development. Companies have become much more successful at creating new markets and promoting lifestyle brand images. The domination of a particular 'style', unique to a particular time and culture, has disappeared, allowing designers a much wider choice of inspiration.

New materials, processes and technology

The stimulus to develop **new materials**, processes and technologies may result from:

59

Table 3.21 Smart materials

Material	Description	Uses
Piezo-electric actuators	Small, slim electronic components that produce a sound in response to an electrical input	Used in novelty greeting cards
'Polymorph'	A polymer that becomes soft and pliable in hot water (62°) and hardens when cool	Used for rapid prototyping of graphic products
Light emitting plastics (LEPs)	Thin, flat, robust, flexible, energy efficient, plastic displays, made by sandwiching a thin layer of polymer between two electrodes. When a low voltage is applied, polymers with different properties can emit red, blue and green light	Likely to be used in hoarding advertisements, safety signage, mobile phones, CD players, TV and computer monitors
Smart ceramic materials	Absorbs and re-emits light energy to 'glow in the dark'	Watch dials, emergency signs, torches
Thermochromatic materials	Microscopic liquid crystal capsules that can be combined with polymers. Changes colour in response to changes in temperature	Kettles, children's feeding spoons, battery test strips

- new legislation, for example, the banning of dangerous materials
- scientific research, for example, the space programme.

Throughout modern design history, newly discovered materials have provided designers with the opportunity to develop products that become desirable because of their 'modern' or 'high-tech' appearance. New processes and technologies have also transformed society because of their functional possibilities. Consider the changes brought about by the silicon chip.

> ## Examiner's Tip
>
> New technologies are emerging all the time and you should keep yourself informed through the media to watch out for new developments.

In recent years, a range of so-called 'smart' materials has been developed. **Smart materials** react to changes in the environment and have led to the development of new types of sensors, actuators and structural components (Table 3.21).

Modern production techniques

Modern materials required the development of new production techniques.

Table 3.22 Examples of modern production techniques

Process	Description
Powder metallurgy	A range of processes including a form of injection moulding used to form pure metal and alloyed components. Increasingly used for precision car components. Small, accurate components for products such as watches. The advantages of this process are that it is accurate and cost effecitve.
Self-chilling aluminium cans	Twisting the can breaks an internal barrier which allows water, trapped in a sealed layer of gel, to vaporise. The vapour is absorbed by a clay-drying agent sealed in the base of the can. This process transfers the heat from the drink into the clay absorber. The advantages of this process are that it removes the need for any refrigeration, and that novelty value attracts customers.

Developments in plastics

The 1950s saw the development of new plastics such as acrylic, PVC and polypropylene, which were perceived as modern, exciting materials.

Table 3.23 Examples of new biopolymers (environmentally friendly plastics)

Biopolymer	Description	Advantages
Enpol	Comparable strength to polythene	Fully biodegradable; two and a half times less material required to achieve comparable performance with conventional plastics
d2w™	A degradable polythene packaging already used by some supermarkets for their carrier bags	Fully biodegradable; can be recycled in the same way as non-degradable plastics
Ecofoam	Made from chips of foamed starch polymer	Ecofoam is water soluble, reusable and free from static; it can replace polystyrene packaging materials

The suitability of plastics for high volume production resulted in a proliferation of products, such as furniture. In the 1970s, the oil crisis and a change in fashion led to a drop in demand for plastics, which became associated with cheap, poor quality and non-ecological products. Plastic's popularity may improve in the future with the development of new eco-friendly biopolymers.

Computers and design

The single most significant technology to influence product development has been the ICT revolution. Computer technology has transformed the way designers work.

- It has enabled the development of small, multifunctional devices, such as wrist watches that include a compass, telephone and navigation utilities.
- CAD/CAM has revolutionised the development of technical and industrial products through the use of, for example, virtual and rapid prototyping.
- The use of CAD graphics has led to new approaches to typography, layout and image making.
- The 'Mac-to-plate' process has revolutionised the digital press and digital printing is increasingly used in fabric printing.
- The growth in home computing has stimulated awareness in the design of graphic products.
- The design of web pages has provided a new medium for designers.

Miniaturisation

Smaller products require fewer materials, less energy and less space. Advances in digital and microchip technologies have enabled designers to produce ever-smaller products. Micro technologies have been superseded by nano technologies, which can construct working devices on a microscopic level. Components, such as gears, have been produced that are smaller than the diameter of a human hair. Suggested applications of micro and nano technologies include:

- tiny robots that are designed to clear human blood vessels
- the world's smallest silicon gyroscope with no moving parts.

Design and the environment

Designers have become more aware of how their work can affect the natural world and our quality of life. Increasingly, product

Fig. 3.7 A microscopic gear shown next to a fly's leg

specifications will include design requirements that help to protect the environment. This may include:

- the use of renewable materials
- reducing the amount of materials used
- the use of recycled materials
- designing for recycling
- using processes that reduce energy consumption
- using processes that do not produce harmful waste products.

Eco-design takes these issues as a priority, ensuring that all design and manufacturing activities will have the minimum of impact on the environment. Recent products include:

- the wind-up Freeplay flashlight
- pencils made from recycled plastic cups
- electric- and solar-powered vehicles with longer lifespans, reduced weight and emissions. Modern cars are designed so that maintenance is simplified and old components can be recycled.

EXAMINATION QUESTIONS

Example questions and answers.

Q1 a) *The globalisation of the market place means that products must be suitable for consumers in different cultures and countries. Describe **two** issues that the designer might have to consider when designing products that will be marketed internationally.* **(2 marks)**

Acceptable answer

1 Most automotive manufacturers supply vehicles around the world and need to ensure that their designs are **easily adapted to suit legal requirements** of different countries including right- and left-hand drive.

2 Design and marketing needs to take account of **linguistic differences** to avoid brand names that have negative connotations in some countries.

b) *Outline the advantages to 'newly industrialised countries' (NICs) and 'less economically developed countries' (LEDCs) of allowing foreign manufacturing companies to site factories within their borders.* **(5 marks)**

Acceptable answer

The jobs created will lead to **higher employment levels and living standards**. The new industries will require new skills, which will **improve the level of expertise of the local workforce**. Taxes and export duties will provide the government with a **new source of foreign currency, improving the balance of payments**. If the industry is new to the area, it will help to **widen the economic base**. In addition, the introduction of new technology may result in a useful **transfer of technology** into other domestic industries.

Professional designers at work

You need to

☐ understand the relationship between designers and clients, manufacturers, users and society

☐ understand professional practice relating to design management, technology, marketing, business, and ICT

☐ understand the work of professional designers and professional bodies.

KEY TERMS

Check you
understand these terms

design and production team, concurrent manufacturing, product data management (PDM), profit, value for money, moral and ethical values

Further information can be found in *Advanced Design and Technology for Edexcel, Product Design: Resistant Materials Technology*, Unit 3B1, section 2.

KEY POINTS

The relationship between designers and clients, manufacturers, users and society

The role of the designer

Artistic and aesthetic role

Often the most significant selling point of a product is the way it looks and the image it projects. Designers often try to inject 'personality' into brand image and products. Designers need to take into account current and future user and market needs, moral, cultural, social and environmental issues, the competition from other products as well as more basic qualities such as shape, form, colour, pattern and style. Automotive manufacturers, for example, spend millions purely on colour research.

Functional and technical role

Designers need to make decisions about function, purpose, materials, systems, construction and finishing. It is important that designers are up to date with the latest technological developments in these areas.

Economic and marketing role

Designers need to be aware of market conditions and should have a clear understanding of production processes and costs so that innovative and attractive products are developed on budget and at the right price.

Organisational and management role

The designer must be prepared to work within a team. **Concurrent manufacturing** is becoming more common and brings together all the different departments to work simultaneously on product development. **Product data management (PDM)** enables fast and easy communication between design, production, suppliers and clients and results in a faster time to market products that meet customer needs.

The role of design and production teams

It is the responsibility of designers and production teams to develop products that match the quality and price requirements of the target market. In order to achieve this, they need to undertake some or all of the following activities.

Identification of needs and opportunities

- Market research
- Research into materials, processes and technology
- Develop a design brief and specification.

Design (including CAD)

- Generate and develop ideas
- Test and model design proposals against the specification.

Production planning

- Produce working drawings and manufacturing specifications
- Organise main production stages
- Plan production schedule
- Plan resource requirements and cost production
- Plan quality control procedures.

Case study: British Airways' 'Go' airline

British Airways commissioned HHCL to help launch its new airline 'Go', which was designed to compete with other increasingly successful low-cost airlines.

Fig. 3.8 The 'Go' brand image

Identifying the need and opportunity

The client/agency project team was briefed to identify the meaning of the brand to the consumer and to develop a complete corporate image. Market research discovered that modern travellers wanted something in between the quality national airlines and low-cost airlines. The 'Go' brand was developed.

Design development

The design agency, Wolff Ollins, was briefed to develop a strong, simple, clever corporate image that would appeal to the target market. The result was the simple and uniquely recognisable 'coloured circles' design. Everything from the corporate stationery to the plane's livery was re-branded.

The success of the product

A simple and manageable advertising campaign was then planned, which resulted in a smooth and successful product launch. The campaign utilised everything from television to sandwich bags to ensure that the target market recognised and understood the brand. The product was so successful that its rival, Easyjet, bought out the company.

Examiner's Tip

You will find it helpful to be able to refer to your own case studies that look at the development of new products or graphic identities and concentrate upon the role of the designer and production teams.

Professional practice relating to design management, technology, marketing, business and ICT

Design and marketing

Good design is not enough to make a successful product; many good products have failed to sell or even reach the market. Marketing involves:

- developing a product marketing plan aimed at the target market group
- providing well-designed, reliable, high quality products at a price customers can afford
- establishing the right image ('lifestyle marketing')
- advertising and promotion (retailers, newspapers, magazines, TV, radio, film, Internet).

Target market groups (TMGs)

Markets are divided into segments ,which classify potential customers according to indicators such as age, disposable income, lifestyle and product end-use.

Marketing plan

A successful marketing plan uses market research to find out:

- consumer needs and consumer demand
- the age, income, size and location of the TMG
- the product type customers want
- the price range they are prepared to pay
- trends affecting the market
- competitors' products and marketing style
- deadlines for the product launch (such as Christmas).

Efficient design, manufacture and profit

Successfully managed product development and production planning will reduce costs and increase profits.

- Careful design management is important as, on average, 80 per cent of costs are incurred from the design stage.
- The difference between the selling price of a product and the costs (design, manufacturing, distributing and marketing) is **profit**.

Efficient manufacture is essential to make a profit. This may be channelled into research and development of new or improved products.

- Production levels have to be planned so that supply will match demand. Unused production capacity is a waste of resources, while insufficient production capacity will result in dissatisfied customers.
- Efficiency is measured as a percentage.

Design for manufacture

Well-designed products will take account of the manufacturing processes. Many companies will analyse design proposals with the aim of making any modifications that will reduce manufacturing costs. These modifications may include:

- simpler designs with fewer components, reducing assembly costs
- changing materials to lighter or more cost-effective alternatives
- choosing materials and processes that reduce waste
- choosing materials and processes that require less energy
- changing the shape of components to make them more suitable for moulding
- altering machining processes to reduce waste or to save time.

Aesthetics, quality and value for money

Designing always involves a compromise between function, appearance (aesthetics), materials and cost. Quality and cost can vary in different products but must match customer expectations in terms of aesthetics and function so that they feel that they are getting **value for money**.

Values issues related to design

The designer has to be aware that his or her **moral and ethical values** might conflict with those of the client, customer or society in general. These include:

- environmental issues (recycling, pollution)
- social issues (affordability, effects on quality of life)
- cultural issues (choices of colour, brand names)
- economic issues (effects of production and marketing on local economies).

Examiner's Tips

- Make sure you can refer to products that illustrate your answer. The clockwork radio, for example, was developed by Trevor Baylis to give poor African communities access to information, such as important health advice. Production was sited in South Africa providing welcomed employment to many disabled people. However, production has been moved to China in order to reduce costs.
- Remember to read around the subject of professional designers at work. Pick a selection of significant designers, working in different fields, which appeal to you. Find out about their design philosophies and look at their work so that you can use the information to illustrate your answers in the examination.

The work of professional designers and professional bodies

Here is a list of some well-known designers and their work:

Product Design

Ron Arad:

- Born in 1951
- Founded One-off Ltd, furniture company
- Designed the Rover Chair (1981)
- Is interested in industrial decay, which is reflected in his one-off furniture products which are constructed from reclaimed industrial materials and components, such as sheet steel and concrete.

Tom Dixon:

- Born in 1959
- Founded SPACE, a shop selling furniture including the Eurolounge range
- Appointed head of design at Habitat in 1998
- Designed the 'S' chair (1987) produced in volume by the Italian company Cappelini

- Inspired by the punk movement, Dixon worked with reclaimed and industrial materials, such as sheet metal and concrete, to produce innovative furniture.

James Dyson:

- Born in 1947
- One of the UK's best-known inventors/designers
- Founded the Dyson company in 1992
- Launched the dual cyclone vacuum cleaner in 1993
- Dyson is synonymous with vacuum cleaners that combine industrial cyclone technologies with boldly styled design. He is also responsible for many other inventions from wheelbarrows to marine vehicles and, more recently, washing machines. He is keen to blur the distinction between designers and engineers.

Michael Graves:

- Born in 1934
- Architect and product designer
- Graves' early work was inspired by classicism and cubism
- He is a postmodernist and has designed in the **Memphis** style for Alessi.

Jasper Morrison

- Born in 1959
- Designed the Universal range for the Italian company Cappelini
- A furniture designer committed to simple, practical and well-known products, Morrison is well-known for his storage units and minimalist furniture using unfinished plywood.

Philippe Starck:

- Born in 1949
- Described as a 'super designer'
- In the 1980s Starck designed the interior of the Parisian café 'Café Costes', which brought Starck to prominence
- Starck designed all the fittings for 'Café Costes', including a three-legged chair
- Likes to design relatively inexpensive products for mass production including the

iconic 'Juicy Salif Lemon Press', designed for Alessi

- Other work includes furniture, televisions, lights, water taps and toothbrushes
- His work fuses a range of influences and enjoys using unconventional combinations of materials.

Architecture and product design

Ettore Sottsass:

- Born in 1917
- One of Italy's best-known designers
- Formed the Memphis design group in 1981
- Sottsass is best known for producing witty and colourful designs influenced by his interest in 1960s Pop Art and Aztec art
- He combined plastic laminates and wood veneers to produce products which contrasted with the modernist style and which was compared with children's toy design.

Graphic Design

Nevil Brody:

- Born in 1957
- One of the UK's best-known graphic designers
- Founded FontWorks UK, an innovative typeface design company (1990)
- Neville Brody designed record covers before becoming a magazine art editor
- His work for *The Face* magazine was highly influential and Brody often ignored the conventions of layout grids and typesetting, influenced by Dada, Russian Constructivism, the Bauhaus and De Stijl.

Peter Saville:

- Born in 1955
- Founding partner of Factory Records
- Has created the visual identity of brands such as New Order, Joy Division, Ultravox, OMD, Wham!, George Michael, Pulp and Suede. His album covers are well known and use computer technology to combine and transform 'borrowed' images with symbolism.

Corporate identity

Michael Wolff and Wally Olins:

- Wolff Olins' company founded in 1965
- Created a new corporate image for British Telecom
- Wolff Olins pioneered the development of innovative corporate identity programmes for companies such as ICI, P&O, Prudential and British Telecom. Changes within BT resulting from privatisation and increased international activity, along with a negative domestic public image signalled the need for a new corporate makeover. A completely new corporate identity was developed by Wolff Olins in response, which included completely new logos, stationery bills, telephone books, vans, shops, offices, company sign systems, telephone boxes, staff uniforms and internal newsletters.

Animation

Nick Park:

- Born in 1958
- Co-Director of Aardman Animation, an animation company
- Created Morph – an early 1970s TV character; Wallace and Gromit – TV characters featuring in a short TV series; Chicken Run – first Aardman animated feature film
- Films start life as storyboards which are turned into 2D, 3D or computer-generated animations. Aardman specialise in 3D stop-frame animation which relied upon detailed, scaled sets and moveable chracacters constructed from a jointed armature, a resin body and modelling clay. Lighting and sound are added to increase the sense of realism
- Successful merchandising (spin-off products) has helped to make the company particularly successful.

The Crafts Council

The Crafts Council was created in 1975 from the Crafts Advisory Committee. It promotes contemporary crafts in the UK and provides services to crafts people and the general public including educational activities, a reference library, a register of designer/makers and a picture gallery.

The Design Council

The Design Council was established in 1960 following the Council for Industrial Design. The aim of the Design Council is 'to inspire the best use of design by the UK, in the world context, to improve prosperity and well being'. The Council achieves this by encouraging business, education and government to work together productively and to communicate more effectively. It provides resources for schools and colleges as well as offering design and marketing advice to professionals.

The Engineering Council (EC)

The Engineering Council was created in 2002 as the main professional body for engineers, technologists and technicians. It is also promotes engineering and technology education through the Neighbourhood Engineer Scheme and Women into Science and Technology (WISE).

EXAMINATION QUESTION

Example question and answer.

 Q1 *The role of the designer and the production team can be summarised under the following headings:*

- *artistic and aesthetic*
- *functional and technical*
- *economic and marketing*
- *organisational and management.*

State the importance of these roles and explain what they might involve. **(12 marks)**

Acceptable answer
Artistic and aesthetic role
In most cases, **a product will not sell unless it looks attractive**. It is the role of the professional designer to determine the look of a product. In order to create an aesthetically successful product,

the designer **must consider qualities such as shape, form, colour, pattern and style.** In addition, the designer **must take account of wider issues such as value issues, competition from other products and future user and market needs.**

Functional and technical role
It is vital that products are **capable of performing the task for which they were designed.** Designers need to make **decisions about function, purpose, materials, systems, construction and finishing.** In order to design successful products, designers **need to be aware of the latest technological developments** in these areas.

Economic and marketing role
The designer must be **aware of economic and market conditions so that the product represents value for money to the consumer.** This means that the designer

should have a clear **understanding of production processes and costs so that price can be kept to a minimum.** In addition, the designer must be **aware of current trends and styles so that the product will appeal to the target market.**

Organisational and management role
Delays cost money and the complex process of product development must be organised and managed effectively for products to reach profitability. Within a **concurrent manufacturing environment, departments work simultaneously on product development and require careful co-ordination. Concurrent manufacturing** systems will often take advantage of **product data management (PDM), which is used for fast and easy communication between design, production, suppliers and clients** and results in a faster time to market of products that meet customer needs.

Anthropometrics and ergonomics

You need to

☐ understand the basic principles and applications of anthropometrics and ergonomics
☐ know about British and international standards.

KEY TERMS
Check you
understand these terms

anthropometric data, standard sizes and dimensions, ergonomics, British Standards Institution (BSI), International Organisation for Standardisation (ISO)

Further information can be found in *Advanced Design and Technology for Edexcel, Product Design: Resistant Materials Technology*, Unit 3B1, section 3.

KEY POINTS

The basic principles and applications of anthropometrics and ergonomics

Anthropometrics

Anthropometrics is the study of human physical dimensions. These include height, width, length of reach, force exerted. Commonly, **anthropometric data** constitutes measurements taken from 90 per cent of the population, that is, between the fifth and ninety-fifth percentile range ignoring the top and bottom five per cent. Designers use this information to design products suitable for this range of people. Data, providing **standard sizes and dimensions**, can be taken from published tables, but it is sometimes necessary to take your own measurements when designing for a person with special needs, for example.

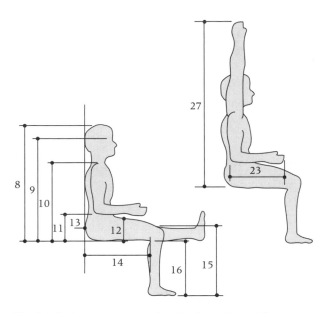

Fig. 3.9 Anthropometric data can be taken from tables and diagrams published for designers

Ergonomics

Ergonomics is the science of designing products for human use. Ergonomics uses and applies anthropometric data to ensure that products and environments are straightforward, safe and comfortable to use.

Interacting with products

Almost all products need to be designed with ergonomics in mind.

- Jewellery needs to be designed to fit the people who wear it.
- Furniture needs to be designed carefully to ensure comfort and safety. When chairs, such as computer chairs, are used for long periods, additional features are incorporated into the design such as adjustable seating positions, foot rests, rollers and contoured cushioning.
- Some products need to be tailored to very specific markets, such as talking calculators for the blind.

When designing products, designers need to consider areas such as shape, form, ease of use, size, weight, colour, noise, materials, maintenance, safety, texture and feel.

Interacting with users

The way in which a person uses a product is also an important ergonomic consideration.

- The size of the hand and the force of the grip are important factors in the design of handles for cutlery, tools and doors.
- Carrying handles must be sufficiently wide enough to allow a large hand to hold them.

Interacting with equipment

Control switches on equipment and machines have to be designed so that they can be operated easily and safely.

- Emergency switches need to be accessible but protected from accidental operation.
- Instrument displays need to be clear and unaffected by reflected light.
- Push button and catch-operated release mechanisms should allow ease of maintenance on products such as vacuum cleaners.
- Weight is an important consideration, especially when designing portable or specialist sporting equipment.

Interacting with environments

Environments include buildings, landscaped gardens, workspaces and vehicle interiors. When designing environments, designers need to consider areas such as movement, light, smell, noise, temperature, space, visibility, facilities, maintenance, safety, furniture and fittings.

- In vehicles, the seating and driving position are adjustable. The driver is able to see and operate all the controls, display gauges and meters. Foot pedals are of an appropriate size, allowing the driver to exert sufficient force. Some systems are power assisted for ease of use.
- Checkouts in shops are arranged so that the lifting is minimised. Barcode readers are both fixed and portable for efficient operation. Prices are displayed for both the operator and the customer.
- Power station control rooms and aeroplane cockpits allow operators to monitor data easily, using gauges and displays. The layout of these must be clear so that they can be located and identified quickly when needed.

British and international standards

British Standards Institution (BSI)

Formed in 1901, the **BSI** is now the world's leading standards and quality services organisation. The BSI works with manufacturing and service industries to develop British, European and international standards. The BSI is:

- independent of government, industry and trade associations
- non-profit making
- recognised globally, operating in more than 90 countries
- serves both the private and public sectors.

The BSI, along with European and international standards organisations, establish national and international standards, testing procedures and quality assurance processes. Products are allowed to display the appropriate logo, such as the BSI 'Kitemark', as long as:

- they meet these standards and testing procedures
- quality systems are in place to ensure that all future products will conform to the same standards.

International Organisation for Standardisation (ISO)

Other international organisations include:

- **International Organisation for Standardisation** (**ISO**): the umbrella organisation responsible for the harmonisation of standards at an international level
- European Committee for Standardisation (CEN), which implements the voluntary

Fig. 3.10 Quality control marks on product packaging. Left to right: the kitemark logo, the 'e' mark, the 'CE' logo

technical harmonisation of standards in Europe
- International Electrotechnical Commission (IEC).

Setting standards

Most standards are set at the request of industry or government to implement legislation. Standards organisations establish safety and product specifications, testing procedures and quality assurance techniques. Products that meet the appropriate standards can carry the logos, such as the Kitemark, but companies need to demonstrate that quality control systems are in place to ensure that products will continue to meet these standards. Such standards help consumers identify safe and quality products and give manufacturers clearly defined standards to produce to.

Ergonomic considerations for designs and models

Designing products involves certain important considerations.

- It is not sufficient to design for yourself or for the 'average' person as this may exclude most of the population.
- Seemingly insignificant errors in design can lead to problems, such as back pain, over time.
- Good aesthetics can mask bad design.

EXAMINATION QUESTIONS

Example questions and answers.

Q1 a) Name **one** national or international standards organisation. **(1 mark)**

Acceptable answer
British Standards Institution (BSI) or **International Organisation for Standardisation (ISO)**.

b) Outline the role of these national and international standards organisations. **(5 marks)**

Acceptable answer

Most standards are set at the **request of industry or government**. Standards organisations establish **safety and product** specifications, **testing procedures** or **quality assurance techniques**. Products that meet the appropriate standards can carry the **logos, such as the Kitemark**, but companies need to demonstrate that **quality control systems** are in place to ensure that products will continue to meet these standards.

PRACTICE EXAMINATION STYLE QUESTION

1 a) Compare the personalities, history, philosophies and design styles of **two** of the following design movements:
   - Arts and Crafts
   - Art Nouveau
   - Modernism and the Bauhaus
   - Art Deco
   - Design of the 1950s and 1960s
   - The New Design. **(10)**

   b) Good design is not enough to make a successful product; many good products have failed to sell or even reach the market. Marketing is just as important as design to the success of a product. Explain the main features of marketing. **(5)**

2 a) Explain the terms:
   - anthropometrics
   - ergonomics. **(2)**

   b) Explain how anthropometric data and ergonomics might be used by the designer in **one** of the contexts listed below:
   - products
   - equipment
   - environments. **(5)**

   c) Explain how design and manufacturing standards are created by national and international standards organisations. **(4)**

   d) Describe how the establishment of standards benefits both the consumer and manufacturer. **(4)**

Total for this question paper: 30 marks

3 B₂ CAD/CAM (R3U3)

This option will be assessed in section B during the 1½ hour, Unit 3 examination. If you have chosen this option, you should spend half of your time (45 minutes) answering all of the questions in this section. It is important to use appropriate specialist and technical language in the exam, along with accurate spelling, punctuation and grammar. Where appropriate, you should also use clear, annotated sketches to explain your answer. *You do not have to study this chapter if you are taking the Design and Technology in Society or the Mechanisms, Energy and Electronics option.*

The impact of CAD/CAM on industry

You need to

☐ understand changes in production methods
☐ understand global manufacturing
☐ understand employment issues
☐ understand the trend from the use of manual CNC programming to the use of software programs that generate CNC codes from drawings
☐ understand the use of software applications that process production data and control a network of different machines from a central system.

KEY TERMS

Check you
understand these terms

CAD, CAM, CIM, EDM, photo realistic images, virtual products, RPT, global manufacturing, remote manufacturing, CNC, G&M codes, FMS, TQM

Further information can be found in *Advanced Design and Technology for Edexcel, Product Design: Resistant Materials Technology*, Unit 3B2, section 1.

KEY POINTS

Changes in production methods

CAD/CAM and product development

CAD (computer-aided design) is used to create, develop, communicate and record design information. **CAM (computer-aided manufacture)** is used to translate design information into manufacturing information including production/process planning. CIM (computer integrated manufacture) is a means of integrating CAD and CAM.

Most products need to undergo a continual process of development through improvements in design or improvements in production methods. The stimulus for change is 'pushed' by the development of new technologies or 'pulled' by the changing demands of the market. Companies need to develop competitive products and services and ICT has become an increasingly important part of the process. Production systems are designed to ensure that the correct personnel, hardware, software, equipment, processes and systems are used to ensure that the client or customer receives the optimum product or service.

The pressures for change

New products are developed for many reasons.

- The emergence of new technologies allows designers to develop new, improved or cheaper products.
- Fluctuating market demand encourages companies to invest in the development of new, improved products.
- The development of new materials, such as smart materials, provides new design opportunities.
- Differences in national legal and cultural requirements lead to the development of design variants.
- Changes in the requirements for product lifespan can lead to longer- or shorter-lived products.
- Political and social changes will lead to changes in the demand for products.

Electronic document management (EDM)

EDM is a means of organising the vast amount of paperwork generated by the modern company. This is achieved by switching from a reliance upon paper-based documentation and correspondence to an ICT-based system, which is much swifter, more efficient and flexible, and more cost effective. Architects, for example, produce a large quantity of plans, which are difficult to manage because of their size. Bureau services allow these companies to convert these into digital files, which allows the architects to archive all their work.

Changes in design and production methods

CAD applications allow the modern designer to:

- develop design concepts and ideas rapidly, down to the smallest detail
- integrate components in an assembly, all drawn to standard conventions
- use CAD applications to establish design dependencies so that changes to values in one part of the design cause changes in all of the dependent values. This is known as parametric designing
- produce **photo realistic images**; the designer can select materials, lighting conditions and lens settings
- produce **virtual products**; three-dimensional images of products can be viewed and manipulated onscreen. Specialist rendering software is used to heighten the sense of realism
- develop design concepts, models and test ideas. Accurate models can be produced quickly using technologies such as rapid prototyping (**RPT**), allowing designers to reduce potential communication problems and to find costly, potential errors or technical and tooling problems
- use DTP to allow designers to carry out a wide range of complex tasks, replacing time-consuming manual processes. DTP also allows designers to output their publications onto film or even directly to plate
- use powerful specialist tools. 'PowerSHAPE' allows designers to take CAD designs and prepare them for manufacture by adding features such as fillets and drafts.

Global manufacturing

The developments described so far, combined with the wider availability of cost-effective transport, have made it possible for people in different areas of the world to work together effectively. Increasing international competition has encouraged companies to respond by looking beyond national borders to find specialist goods and services or to reduce costs. Large engineering projects commonly draw parts and utilise services from many different countries and many large manufacturers are able to transfer production to take advantage of lower wage costs and other incentives. This is known as **global manufacturing**.

Remote manufacturing

The ability to communicate instantly, using video conferencing, with people all over the world, combined with new technologies allowing reliable electronic data exchange, means that designs can be manufactured anywhere in the world. This is known as **remote manufacturing**.

- Video conferencing and other communication technologies make sure that designs can be developed in discussion with the manufacturer, ensuring they meet the constraints of the manufacturing process.

- The finished design can be sent electronically directly to the manufacturing centre where it is machined using **CNC** equipment. The process can be monitored by the design team using video links.
- The whole process is very quick. The finished components can be checked and dispatched the same day.
- The whole process allows designers to take advantage of CNC manufacturing technology without having to make the heavy investment necessary to purchase all the machinery.

Employment issues

CAD/CAM technologies and the growth in new working practices have led to changes in the patterns of employment. Increased automation has led to a reduced labour force. New technologies have created a demand for new skills.

Employment trends within design industries

Traditionally, designs were hand drawn by teams of skilled draughtsmen who worked together in large drawing offices. The development of CAD and related technologies has led to significant changes in working practices.

- Designers need to be highly skilled and more flexible in their approach.
- Designers are no longer tied to a location.
- Skills need to be updated regularly as new generations of software are developed.
- Smaller manufacturers may not be able to employ adequate numbers of these skilled workers and so turn to specialised bureau services to provide CAD/CAM services.

Employment trends within manufacturing industries

Traditionally, the skilled and semi-skilled workforce grouped itself into narrowly specialised trades. Demarcation placed rigid barriers between workplace activities, which led to an inflexible workforce. New CAD/CAM technologies have created a demand for new skills and models of workforce organisation.

- Workers need a broader range of skills.
- Ongoing training is a feature of these new professions.
- All employees need to be more flexible in their approach, be prepared to take on new responsibilities, and to operate as part of a team, cell or work centre. Workers and resources are directed to where they are needed to reduce bottlenecks and production times.
- Job security has weakened as manufacturing companies are more prepared to relocate to take advantage of fluctuating markets and changing economic conditions.
- These new demands have often led to skill shortages.

Future development of CAD/CAM technologies

Systems that rely on ICT, including CAD and CAM, must evolve to remain competitive. There is always pressure to increase productivity and to reduce the 'lead time' from product development to market. This requires continuous research and development into data manipulation and communication systems. Developments in the use of DTP, electronic data transfer and mobile communications are examples of evolving technologies that have helped to increase productivity. Innovation can be categorised into three areas.

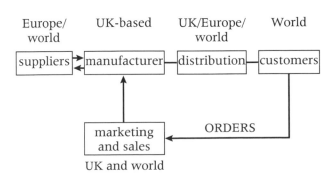

Fig. 3.11 The worldwide distribution of a UK manufacturing company

- Critical technologies: the 'building blocks' from which products develop. Innovation arises from the development of new sensing and control systems, materials handling, storage and retrieval systems and the development of robotics technology.
- Enabling technologies: critical technologies are used to develop new products such as CNC machinery.
- Strategic technologies: these are concerned with planning systems that incorporate critical and enabling technologies. Decisions have to be made concerning capital investment in new products and factory layout.

Computer numerically controlled (CNC) machines

CNC machines are controlled by **G&M codes**, which are a list of number values and co-ordinates. Each number or code is assigned to a particular operation. These used to be typed in manually by machine operators. Nowadays, most CAD software is able to generate these G&M codes automatically from the design drawings. The advantages of CNC machines are shown in Table 3.26 later in this unit.

Computer integrated manufacture (CIM)

Traditional linear approaches to manufacturing ('over the wall') are straightforward but can have major drawbacks; design errors and manufacturing problems take longer to identify and 'lead times' (the time taken to develop a product so it is ready for sale) are much longer.

Concurrent engineering

CIM systems enable concurrent engineering where multidisciplinary teams start working together from the start of product development. Errors and problems are recognised much more quickly and lead times are reduced.

Flexible manufacturing systems (FMS)

Thorough, responsive planning and organisation are essential in the modern business in order to respond quickly to changes in demand and customer requirements. The features of a **FMS** include:

- multipurpose, automated ICT and CNC equipment
- centralised control of a network of systems and machines
- robot technology
- relational databases making data accessible
- computer-aided process control (CAPP), used to plan manufacturing operations
- computer-aided production management (CAPM) uses data to manage manufacturing operations.

It is much easier to make last minute changes and to organise the most efficient use of resources without compromising on quality.

Total quality management (TQM)

The improved management of design and production data allows quality to be monitored closely through all stages of product development. Total quality management (**TQM**) seeks to develop a culture where all employees are responsible for ensuring quality and are committed to continuous improvement (CI).

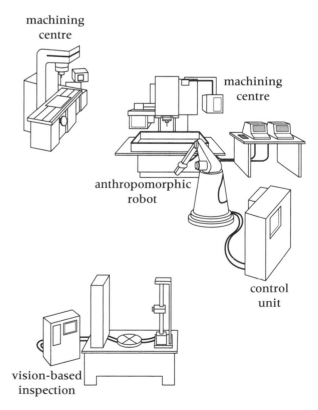

Fig. 3.12 A flexible manufacturing cell

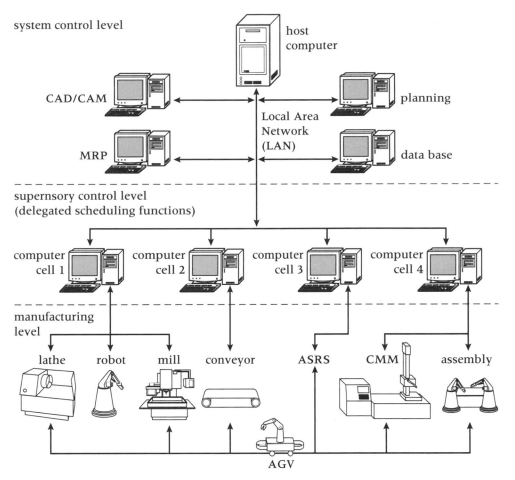

system control level

host computer

CAD/CAM

planning

Local Area Network (LAN)

MRP

data base

supernsory control level
(delegated scheduling functions)

computer cell 1

computer cell 2

computer cell 3

computer cell 4

manufacturing level

lathe   robot   mill   conveyor   ASRS   CMM   assembly

AGV

Fig. 3.13 A schematic diagram of a centrally controlled CIM system

The use of software applications that process production data and control networks

The advantages of centralising and integrating control systems has been discussed above. CIM and FMS systems, for example, rely on sophisticated software to control and co-ordinate a complex range of processes and CNC machinery. Figure 3.13 illustrates the range of centrally controlled operations within a CIM system.

EXAMINATION QUESTIONS

Example questions and answers.

a) *Give **three** advantages of computer integrated manufacture (CIM).* **(3 marks)**

Acceptable answer

1. CIM allows a **number of people to work on a project** at the **same time** (concurrent engineering).
2. CIM **increases the effectiveness of quality control using automated sensing systems**.
3. CIM **increases flexibility** by allowing **changes to designs or production processes** to be made **quickly**.

b) *Describe the consequences for the workforce of increased reliance on CAD/CAM technologies.* **(4 marks)**

Acceptable answer

Workers need a **broader range of skills**, which require an **ongoing training programme** so that they can keep up to date with developing technologies. These new demands have often led to **skill shortages**. Employees cannot rely on a job for life as **job security has weakened** as

manufacturing **companies are more prepared to relocate** to take advantage of fluctuating markets and changing economic conditions.

Computer-aided design

You need to

☐ understand the use of CAD to aid the design process
☐ know common input devices
☐ know common output devices.

KEY TERMS

Check you understand these terms

PCBs, virtual products, total design concepts, multimedia, input devices, output devices

Further information can be found in *Advanced Design and Technology for Edexcel, Product Design: Resistant Materials Technology*, Unit 3B2, section 2.

KEY POINTS

The use of CAD to aid the design process

Components of a CAD System

A computer-aided design (CAD) system is the combination of software and hardware used to create, develop, test, communicate and record design information.

Hardware

CAD programs require powerful processors (CPUs) and lots of memory (RAM). CAD systems used to require specially designed (dedicated) systems and mainframe computers/minicomputers but, due to advances in the power of personal (micro) computers

(PCs), it is possible to use this software on conventional networks (LANS, Intranets and WANS) or on stand-alone computers.

Software

Hardware is useless without software. Operating systems (OS), such as Windows or Unix, allow several tasks or programs to run at the same time and provides a graphical user interface (GUI), which offers a more user-friendly method of communication through the use of menus, buttons and icons or through keyboard 'shortcuts'. Modern software also allows users to customise their workspace. Applications, such as CAD programs, work within the OS environment and contain hundreds of functions, enabling the user to accomplish specific drawing tasks.

Software standards

To ensure compatibility, it is important that designers and engineers can work to common standards. Designers need to work to standards such as ANSI, ISO or BSI for technical

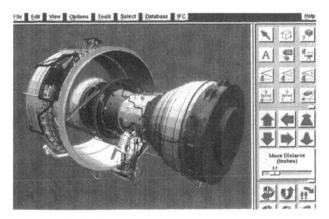

Fig. 3.14 CAD/CAM is used extensively to create, develop, test, communicate and record design information

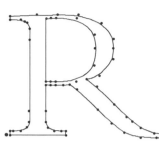

Fig. 3.15a Vector graphics: stored as a line drawing with colour fills. Can be manipulated in many ways without losing image quality. Small sized files

Fig. 3.15b Raster graphics: (bitmaps) stored as a large collection of different coloured pixels (tiny squares). Limited manipulation possible but enlarging leads to loss of image quality. Large sized files

engineering drawings. Some standards have been established by recognised standards organisations, but many CAD/CAM standards have tended to be established over time by the most successful companies in any particular field. These include:

- Hayes command set for modems
- Hewlett Packard Printer Control Language (PCL) for laser printers
- Postscript (PS) page description language for laser printers
- Data Exchange File (DXF) for CAD files created by AutoDesk.

Managing CAD data

CAD data can be stored in many forms including hard discs, magnetic tape, CD(R)-ROM, and network servers. Large files can be compressed using WinZip or similar programs. There are many file formats including DXF, WMF, PICT, and VRM.

Constructing accurate drawings and complex products

Engineers use CAD in a variety of ways to design and test a product's function, form, structure and aesthetics. Materials, surface finishes and dimensions can be specified within the drawing. Modern CIM systems are also able to share design data among different departments, suppliers and systems. Applications include:

- promotional images to sell the product
- sales presentations and company reports
- marketing materials and brochures for company websites
- system flow diagrams

- PCB track design
- accurate and complex drawings that contain all the information necessary to make the product
- drawings used to test the product.

2D and 3D drawings

Traditionally, 2D drawings were used to communicate designs. They are still important but the advantages of 3D software now mean that most products are developed in 3D. Most 3D CAD applications will generate 2D drawings automatically.

CAD drawings must communicate all the information required to manufacture the design. Complex products can be represented clearly and accurately using a wide range of features offered by a professional CAD system. There are two types of drawings:

- *detail drawings*: representations of individual parts
- *assembly drawings*: a representation of all parts that go together to form a product.

In order to communicate all the information required, designers may use a range of drawing conventions.

- Sectional views allow the designer to slice the product open along a 'cutting plane' to show hidden features. Cross hatching shows cut surfaces. CAD programs can generate these views automatically.
- Exploded views show the product partially disassembled (exploded) and are used to show assembly details and internal features.

Dimensioning and annotation

It is essential to provide accurate dimensions and additional details to support CAD generated drawings. In industries that take advantage of global manufacturing opportunities, it is essential that everyone uses the same conventions and works to the same standards. CAD programs generate dimensions automatically and can be set to use ISO standards to ensure the correct practices are used. General principles of dimensioning include:

- the minimum number of dimensions should be used
- dimensions should incorporate appropriate tolerances
- dimensions should not interfere with the drawing wherever possible
- dimension lines should not cross.

CAD modelling

It is far more cost effective to make mistakes and modifications at the early stages of product development than later in the process. CAD allows products to be modelled digitally so that virtual prototypes can be modified and tested onscreen, reducing the expense of producing actual prototypes. Commonly used functions, such as sweep techniques, are incorporated as commands in drop-down menus or as buttons. The software requires powerful computers and processors to carry out the complex mathematical calculations.

- *Wireframe models* are 'transparent' representations of the product constructed from lines. Surfaces are not shown and cannot therefore be rendered. They are useful for showing external features and details but can be difficult to understand.
- *Surface models* are 'translucent' representations that share some of the advantages of wireframe models but which are easier to understand.
- *Solid models* are 'solid', opaque representations of the product. These are the clearest representations and they also contain the most data about the product.

Rendering

Rendering is the process of applying colour, pattern, texture, light and shade to a computer model. Realism is improved by hiding unwanted elements, such as construction lines, and adding backgrounds. Solid, rendered models are opaque, so hidden details remain hidden. CAD and specialist rendering software use a range of methods to render onscreen objects including:

- flat shading
- graduated shading
- phong shading (slower process that generates highlights to improve realism)
- texture mapping (slow, memory intensive process that applies 2D textures around 3D forms. Textures apply 2D properties such as colour and brightness and 3D properties such as transparency and reflectivity).

Creating and modifying designs and layouts

Computer to plate and digital technologies

Traditional methods of producing printing plates involve the creation of negative or positive films from camera-ready artwork. The film is then used to expose photosensitive printing plates, which then have to be developed. 'Computer to plate' techniques use laser technology to produce printing plates directly from digital files. This shortens the whole process, reduces costs and improves the quality and definition of the end product by

Fig. 3.16 Crash Bandicoot – A three-dimensional computer-generated character from a computer game

cutting out the intermediate stages of production. Modern digital printers are now able to print high quality images onto a range of materials and products. These printers can be used to produce high quality mock-ups for new product lines or to personalise marketing materials, such as mouse mats.

Using computers to design PCBs

Printed circuit boards (**PCB**s) are used to link components to form an electronic circuit. Designers rely on computer technology to test virtual circuits and to plan the route of the copper tracks on the PCB. The design data is then passed to the production engineers who use it to set up the CNC machines used to automatically insert the components. There are three stages.

1 The schematic circuit is drawn as a diagram using symbols.
2 The circuit is simulated to test for faults.
3 The track layout is auto-routed.

Creating virtual products and total design concepts

Virtual reality programs enable designers to generate virtual designs that allow the user to interact with the objects or environments on the screen. Architects can construct virtual buildings, which allow the client to 'walk' through the building, viewing it from different angles or in different weather conditions. **Virtual products** can be created that allow the customer to view the object from different angles.

It will be much easier to develop and communicate design proposals and **total design concepts** in the future with the help of ICT. Two dimensional and three dimensional CAD images, animations, text, sound and global communications technology can be combined together to create a detailed and comprehensive **multimedia** working environment. Designers will be able to work together with the help of 'expert' systems linked to large databases of production data. Features of these systems will include:

- communications systems integrating video conferencing, intelligent whiteboards and intranets
- rapid data transfer systems, such as broadband, allowing engineers and designers to work together on the product simultaneously
- multimedia presentations for clients and customers, using VRML to show virtual products
- efficient management of communications that automatically routes electronic documents to people who need them.

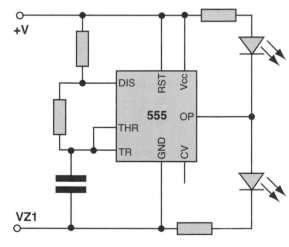

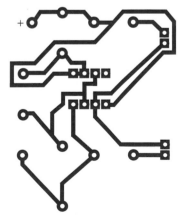

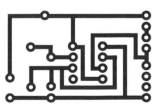

Stage 1
Software can be used to simulate different circuits. Components are shown as symbols in a schematic diagram. The circuits can then be tested on screen and the software will identify problems such as overloaded components.

Stage 2
When the designers are happy with the circuit they can use auto routeing software to generate tracks for PCB.

Stage 3
The auto routeing software will arrange the tracks to make efficient use of space and will output the masks needed to produce the PCB.

Fig. 3.17 A simple circuit designed using simulation and auto routeing software

Common input devices

Input devices describe hardware that is used to control or enter data into a computer. Command-driven interfaces have been largely replaced by graphic user interfaces (GUIs). These GUIs use pull-down menus and icons to generate commands and manipulate the design onscreen.

Input devices tend to produce analogue signals that need to be converted into digital signals by a digital signal processor (DSP). Input devices allow:

- ease of control by clicking on icons
- information to be communicated
- data entry, such as dimensioning
- selection and manipulation of the design.

Table 3.24 Input devices

Hardware	Description	Use	Advantages
Mouse	Mechanical or optical versions often incorporating a scroll wheel. Used to move cursor and to select software functions but not very accurate. Wireless mice also available.	Everyday computer use	Inexpensive; easy to use
Tracker-ball	An 'upside down mouse' with large ball and buttons. Rolling the ball moves the cursor.	Graphic programs, such as VR, where images are constantly being moved	Remains stationary; more precise than mouse
Digitiser	Large reactive electronic table and 'puck'. Pre-programmed buttons operate commonly used operations. Puck is moved over table surface and is used to send co-ordinates to screen or to select functions.	Technical drawings	Very accurate; drawings and sketches can be transferred directly to screen
Graphics tablet	Pressure sensitive tablet and stylus (pen), available in a range of sizes. Program functions can be selected from pre-programmed areas of tablet. Pen is moved over surface of tablet to create an image, which is transmitted to the screen.	Art/design; artwork	Senses stylus pressure, allowing variation in line quality; closest to drawing by hand
Photo CD	CD-ROMs can be used to store large amounts of data required by large, high-resolution image files. Computers with CD drives can be used to manipulate images and to send to printer.	Personal photos; design images	Easily transportable; copies can be printed; saves using space on hard disc
2D scanners	These allow the designer to input analogue data and photographs, creating digital, raster images that can be manipulated using different software applications. Resolution is measured in dots per inch (DPI). Scanning produces large file sizes, especially when using colour. Can be used in conjunction with OCR software. A wide range of scanners are available including handheld, flatbed, sheet-fed, drum, overhead and transparency scanners.	Inputting line artwork to be developed into vector graphics; OCR; to digitise photographs or transparencies	Digitised images can be stored and transmitted more efficiently; a wide range of software allows images to be manipulated onscreen; images can be incorporated into publications or web pages

▶

Table 3.24 Input devices *(continued)*

Hardware	Description	Use	Advantages
3D scanners	Objects can be scanned in three dimensions and the digital data used to create 3D computer models. There are two types of scanners: contact scanners (which use a probe to supply geometric data) and non-contact scanners (e.g. laser, ultrasonic and magnetic scanners, which use triangulation to locate points). Most scanners are available in manual and automatic forms.	To transfer solid concept models into digital form; to make copies of existing forms	Allow designers to produce relatively accurate computer models of solid objects; increase flexibility of design development
Digital video Digital cameras	Digital video cameras can be used to record moving images. Digital cameras do not use conventional film but store images in digital format, which can be down loaded directly into the computer.	Personal use; CD-ROMs; Internet Used to create original digital images	Low operating costs; speed of data conversion; reliance on film/hard copy removed; digitised images can be stored and transmitted more efficiently; a wide range of software allows images to be manipulated onscreen; images can be incorporated into publications or web pages

Common output devices

Output devices describe hardware connected to a computer that converts data or designs into a useful form. CNC technology needs an interface between CAD and CAM to translate the drawing data into a useable form. The connection is made using cables or, in some cases, via wireless infrared or radio technology.

Table 3.25 Output devices

Hardware	Description	Use	Advantages
Dye sublimation printers	Expensive, high quality but slow process. The four colours of ink (dye) are stored on a film, which is transferred onto paper using a heated print head, which turns the ink into a gas. The amount of ink released can be adjusted by varying the temperature of the print head.	Used where the highest quality prints are needed	Very high quality; precise dense colour without dots or dithering; smoothly gradated tones
Monitors	Monochrome, greyscale and colour.	Visual output	Larger, higher resolution monitors (measured by dpi) display more information
Thermo autochrome (TA) printing	Requires special TA paper, which is coated with three layers of coloured pigment, each activated by a different temperature. Three passes are required, each at a different temperature, and at each stage the colour is fixed with UV light.	High quality digital prints	Very high quality

Table 3.25 Output devices *(continued)*

Hardware	Description	Use	Advantages
Digital printing	Digital printing technologies link modern, digital printing presses and computers that can print directly without the need to make printing plates. Based on fast, high quality laser technologies, these printers come with a full range of image handling capabilities, much like some of the tools found in DTP software, such as image cleanup, enlargement and reduction.	Affordable technology for small and large printing companies and reprographic departments in larger companies	Fast (250ppm), high quality output, flexible, centrally controlled; production details can be specified and attached to print files; documents can be seen onscreen as they will be printed; simple to manage multiple print jobs – jobs can be rearranged to avoid unnecessary changes in paper; jobs can be stored easily for repeat orders; more effective response to fluctuating market demands

EXAMINATION QUESTIONS

Example questions and answers.

Q1 a) Name **one** input device that is commonly used with CAD applications. **(1 mark)**

Acceptable answer
Digitiser.

b) Many CAD programs are described as 'parametric'. Explain what this means. **(2 marks)**

Acceptable answer
When using a fully parametric CAD application, the designer or engineer can **make a change to one part of the design**, such as changing a dimension, and **the software will automatically adjust the rest of the design to accommodate the modification.**

c) Describe **three** specialist features of CAD programs (not including parametric features) which help designers and manufacturers. **(6 marks)**

Acceptable answer
1 **Manufacturing instructions can be embedded in the design**, such as attributing line colours for different operations. **This speeds up the manufacturing process as it avoids the need to start and stop CNC machinery to change settings manually.**
2 **Working drawings and parts lists can be generated automatically**, which **speeds up the process and improves the quality of information available to manufacturers.**
3 Much **time can be saved** by utilising the libraries of pre-drawn 'primitives' or pre-drawn components.

Computer-aided manufacture

You need to

☐ understand the applications, advantages and disadvantages of common CNC machines

☐ understand the use of CAM in one-off, batch and high volume/continuous production
☐ understand the advantages and disadvantages of CAM.

KEY TERMS

Check you
understand these terms

RPT, LOM, cutting tools, tool paths, speeds, feeds

Further information can be found in *Advanced Design and Technology for Edexcel, Product Design: Resistant Materials Technology*, Unit 3B2, section 3.

KEY POINTS

The applications, advantages and disadvantages of common CNC machines

Rapid prototyping

CAD/CAM technologies allow the designer to test ideas very quickly through rapid prototyping (**RPT**). Physical models can be generated directly and automatically from onscreen designs using CNC equipment. Stereolithography is a 'tool-less' process that

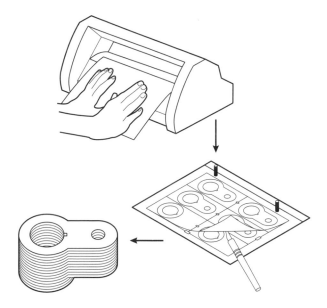

Fig 3.18 Layered object modelling

uses laser technology to solidify liquid polymers to form complex 3D shapes. Layered object modelling (**LOM**) produces models from CAD drawings, made from layers of self-adhesive card, which are built up on a pegged jig.

- Accurate models can be generated very quickly from CAD drawings.

Table 3.26 CNC machines

Hardware	Description	Use
CNC lathes	Automated versions of manual lathes; programmed to change tools automatically	Turning and boring wood, metal and plastic components
CNC routers, milling machines and engravers	3–5 axis versions	Wood, metal and plastic: 3D prototypes, moulds, cutting dies, printing plates, sign making
CNC cutting machines, such as flame, laser, spark erosion cutting machines	A wide range of specialist machines	Cutting and binding printed material
Pressing, punching, bending, and die-cutting machines	A wide range of specialist machines	Used to process sheet metals, plastics, paper and boards: flat carton packs; document wallets
Knitting, sewing, embroidery machines and looms	Computer controlled textile machines	Used to construct or enhance textile products
Printers	Wide range including inkjet, laser, dye sublimation, thermo autochrome and digital printers	Ranging from general purpose printing to specialist applications

- RPT improves ability of designers to communicate designs to colleagues and clients.
- Potential errors, such as tooling problems, can be identified more readily.

Features of CNC machinery

Processing flexibility is the key feature of CIM and CNC machines. Features of CNC equipment include:

- the tool or material moves, or both
- tools can operate in 1–5 axes
- larger machines have a machine control unit (MCU) that manages operations
- movement is controlled by a series of servo-motors or stepper motors (actuators)
- feedback is provided by sensing devices (transducers) or encoders
- tool magazines are available to allow automatic tool changes.

Tools

- Most **cutting tools** are made from high-speed steel (HSS), tungsten carbide or ceramic materials.
- Tools are available in a wide range of profiles determined by their purpose.
- Tools are designed to direct waste material away from the work (e.g. flutes on milling cutter).
- Tools are usually held in collets.
- Some operations require coolant, such as oil, to protect tool and work.

Tool paths, cutting and plotting motions

- **Tool paths** describe the route the cutting tool takes.
- Motion can be described as 'point to point', 'straight cutting' or 'contouring'.
- **Speeds** are the rate at which the tool operates (e.g. rpm of a collet, spindle or chuck).
- **Feeds** are the rate at which the cutting tool and workpiece move in relation to each other (e.g. the rate (mm/s) at which a lathe tool moves 'into' the workpiece).
- Feeds and speeds are determined by the cutting depth, material and quality of finish required. Harder materials, for example, generally require slower speeds and feeds.
- Roughing cuts remove larger amounts of material than finishing cuts.
- Rapid traversing allows the tool or workpiece to move rapidly when no machining is taking place.

Scale of production

The scale or level of production will fall into one of three categories: one-off, batch, high volume or continuous production. You may need to refresh your memory of this subject by referring to Unit 1.

Recent developments

Flexible manufacturing systems rely on ITC to provide 'quick response' systems that can customise products to meet the requirements of the client.

Outsourcing allows smaller manufacturers access to specialist ICT resources in order to provide a real FMS service to clients. CAD bureau services, for example, provide specialist skills and access to expensive resources, which can be used by smaller companies that are unable to make the heavy investment necessary to offer these services in-house.

Distributed numerical control (DNC) 'part programs' can be downloaded into the memory of the MCUs as and when required from a central computer within a CIM system. This avoids tying up the computer, which can continue to service and co-ordinate the rest of production. The stages in part programming are:

1 CAD program identifies required machining operations
2 Appropriate tools are suggested or selected from a tool library
3 Tool paths are calculated
4 Machining is simulated onscreen
5 Errors, such as collisions, are identified and displayed
6 A cutter file is produced, which can be read by the CNC machine
7 The cutter file is transmitted to the CNC machine.

Table 3.27 The advantages and disadvantages of CAM

Advantages of CAM	Disadvantages of CAM
• *Speed*: although CAM machines may seem slow sometimes, they work much faster than humans. • *Accuracy*: even school-based CAM machinery will work to tolerances approaching +/− 1/100th of a millimetre. Higher precision and less human error leads to less waste. It is now possible to manufacture very complex designs. • *Reliability*: the software will check and simulate the machining operation. • *Repeatability*: it is easy to reproduce identical components. • *Productivity*: more products can be produced because CAM machines can work continuously without the need for breaks. • *Safety*: since the operator is not in direct contact with the tools or materials, accidents happen rarely. CAM machines can work with dangerous materials. • *Flexibility*: highly suitable for batch production and JIT where systems need to switch production between jobs regularly. Many machines are multifunctional, such as plotter-cutters. • *Operating costs*: speed and automation of previously manual operations lead to low unit and operating costs. There is also less waste.	• *Employment issues*: reduction in workforce as many manual tasks become automated. Some human operations become repetitive and undemanding as the machines do all the work. • *Skills*: increased flexibility and new skills required from the workforce. • *Investment*: CNC machines are expensive and require a large initial outlay.

EXAMINATION QUESTIONS

Example questions and answers.

Q1. a) *A school has purchased a new minibus and has approached a local sign maker to produce the school name, which will be applied to the side of the minibus. What type of CNC machinery is most likely to be used to produce the school name?* **(1 mark)**

Acceptable answer
(CNC) vinyl cutter/(CNC) plotter-cutter.

b) *The sign making company has invested heavily in CNC equipment. Signs and transport livery used to be produced by hand. Describe **two** negative consequences of making such a change in production methods.* **(4 marks)**

Acceptable answer
Investing in CNC machinery may lead to a **reduction in the workforce** because many **manual tasks become automated**.

CNC machines are **very expensive** compared with the equipment needed to produce signs manually. It will take a **long time** before the company sees a **return on the large initial investment**.

PRACTICE EXAMINATION STYLE QUESTION

1 a) Describe, briefly, how designers might make use of **two** of the following devices. **(4)**
   - mouse
   - tracker-ball
   - digitiser
   - graphics tablet
   - photo CD
   - 2D scanners
   - 3D scanners
   - digital video
   - digital camera.

 b) Describe how specific tools and features within CAD programs aid the designer or engineer. **(5)**

 c) Describe the use of virtual products in the modern design and manufacturing organisation. **(3)**

 d) Explain, briefly, the meaning of the following innovation technologies:
   - critical technologies
   - enabling technologies
   - strategic technologies. **(3)**

2 a) Explain **two** of the following terms.
   - Computer Numerical Control (CNC)
   - G&M codes
   - Computer integrated manufacture (CIM). **(4)**

 b) Name **one** specific example of CNC machinery. **(1)**

 c) Explain the advantages of using CNC machinery during manufacturing. **(5)**

 d) Describe the use of remote manufacturing. **(5)**

Total for this question paper: 30 marks

3 B₃ Mechanisms, energy and electronics (R304)

This option will be assessed in Section B during the 1½ hour, Unit 3 examination. If you have chosen this option you should spend half of your time (45 minutes) answering all of the questions for this section. It is important to use appropriate specialist and technical language in the exam, along with accurate spelling, punctuation and grammar. Where appropriate you should also use clear, annotated sketches to explain your answer. *You do not have to study this chapter if you are taking the Design and Technology in Society option or the CAD/CAM option.*

Mechanical systems

You need to

☐ **analyse mechanical problems**
☐ **add feedback to achieve closed-loop control.**

KEY TERMS
Check you understand these terms

mechanical systems, input, control, output, open-loop, closed-loop, feedback

📖 **Further information can be found in** *Advanced Design and Technology for Edexcel, Product Design: Resistant Materials Technology*, **Unit 3B3, section 1.**

KEY POINTS

Analysing mechanical problems

Mechanical systems are all around us, from turning up the volume on the stereo to turning a tap to get water out. Each system has been designed for a specific purpose but they all follow the same principles and have an **input**, a **control** and an **output**.

System blocks of a control system – a hole punch

The hole punch is a very simple mechanical system in that it has an input, a control mechanism and an output. You press the lever down and the paper is punched. The input in this case is in the form of a movement. It is this which starts, or triggers, the system. The control element is a signal processor. It responds to the input signal and normally

| Apply force to lever | → | Lever mechanism controls punch | → | Holes punched in the paper |

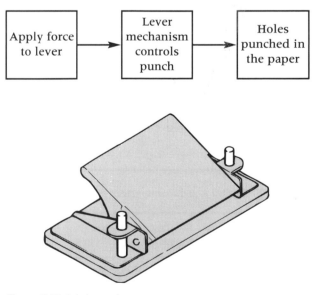

Figure 3.19 A hole punch

changes the size of the input by means of levers, gearboxes or linkages. The output is the end result of the control system. In mechanical systems, the output is in the form of movements.

The hole punch is an **open-loop** control system. This means that the individual blocks are connected one after the other in series, or in a linear fashion. Once the system has been triggered, each block is processed and actions carried out one after the other. In an open-loop system, there is no checking to make sure that the previous block has been completed.

Add feedback to achieve closed-loop control

A **closed-loop** system essentially works in the same way as an open-loop system, with the addition of a feedback loop built in to monitor and check what is happening. This loop also checks whether previous blocks have been carried out successfully. The **feedback** loop is continually monitoring the state of the system and updating the input accordingly. Control systems are widely used in CNC machinery to monitor speed, feed rates and the position and movement of the tool in relation to the datum point.

EXAMINATION QUESTION

Example question and answer.

 Q1 *Give an example of a mechanical system and describe the input, control and output.* **(4 marks)**

Acceptable answer
A **door handle and mechanism** is an example of an open-loop system. The **handle**

acts as a lever and is the input. As the handle is rotated, a **spring-loaded mechanism inside the door fitting converts the rotary motion into a linear movement**. The resulting **output is the drawing back of the barrel from inside the door frame**, which subsequently allows the door to be opened.

Energy sources

You need to

☐ be aware of the principles of generating renewable energy.

 KEY TERMS
Check you understand these terms

capital energy, income energy, renewable

Further information can be found in *Advanced Design and Technology for Edexcel, Product Design: Resistant Materials Technology*, Unit 3B3, section 2.

KEY POINT

The principles of generating renewable energy

Energy can be obtained from direct sunlight, fossil fuel, nuclear, and movement of the water

Table 3.28 Capital and income energy

Capital energy	Forms of energy that cannot be replaced once they have been used, such as gas, oil and coal.
Income energy	Forms of energy that continue to be available and are **renewable**, i.e. they will not run out, such as solar, nuclear, wave, tidal and geothermal.

and air. Energy is categorised into two groups (Table 3.28).

Wave energy

Waves are formed by winds blowing across the surface of the oceans and by tidal patterns. It is quite difficult to harness the power from waves. One method uses floating rafts connected together, causing a mechanical movement as they bob up and down. The mechanical movement is then converted into electricity.

Tidal energy

Tides are caused as a result of the Earth's gravitational pull from the sun and moon. The tides have enormous power and create huge forces; certain areas in the British Isles have tide level changes of up to 15 metres. Energy is harnessed by making the water flow through channels or by the changing volumes of water filling and emptying huge basins. In both instances, turbine blades are made to rotate, which in turn are connected to electrical generators.

Geothermal energy

Geothermal energy is taken from below the Earth's surface by drilling deep holes and pumping water down through them. The water is heated naturally to quite high temperatures before the heat is extracted on the surface. This type of energy extraction is prevalent in Iceland and New Zealand.

Solar energy

Solar energy can be harnessed in several ways: solar panels, cells or furnaces.

- Water pumped in small bore pipes through solar panels absorbs the sun's energy and is then fed into domestic hot water systems.
- Solar cells are also known as *photovoltaic cells*; these convert the sun's energy directly into electricity. Although they are considered to be quite expensive, they have become increasingly efficient in recent years.
- In a solar furnace, mirrors focus the sun's rays onto a focal point. Here, water is heated or, in some instances, food can be cooked.

Fossil energy

Coal, oil and gas are all fossil fuels that have been made naturally over millions of years. They are being used up at ever-increasing rates because of growing energy needs. Once they have been used, they cannot be replaced. These types of fuel are all burned to generate heat in homes. They are also used on an industrial scale to generate electricity: they are burned to heat water, creating steam to drive turbines that generate electricity.

Nuclear energy

Uranium is used to produce heat in nuclear fusion reactors. The heat released is used to heat water, producing steam to drive generators and subsequently to generate electricity.

EXAMINATION QUESTION

Example question and answer.

 Q1 *Energy can be generated from the following sources:*

- *wave*
- *geothermal*
- *nuclear.*

Select **one** *of these sources and briefly describe how energy is generated from it.* **(2 marks)**

Acceptable answer
Waves are channelled through ducts, which causes air to move quickly. The 'rush' of air causes **a turbine blade to rotate, which in turn drives a generator** to create electricity.

Levers and linkages

KEY TERMS
Check you understand these terms

lever, mechanical advantage (MA), velocity ratio (VR), efficiency, linkages

📖 **Further information can be found** in *Advanced Design and Technology for Edexcel, Product Design: Resistant Materials Technology*, Unit 3B3, section 3.

KEY POINTS

Classes of levers and linkages

A **lever** can be defined as a rigid rod that pivots about a fixed axis, known as a *fulcrum*. There are three different types of levers, classified as Class 1, 2 or 3.

Class 1 and 2 levers are the most common because they provide a **mechanical advantage (MA)**.

The mechanical advantage can be easily calculated by comparing the load (L) with the effort (E). A high mechanical advantage means that you can move large loads with a small effort.

Explain and use simple calculations

Taking the example of the wheelbarrow (a Class 2 lever) in Figure 3.20, the MA can simply be calculated using the formula:

$$MA = \frac{load \ (L)}{effort \ (E)} = \frac{400N}{100N} = 4:1 \ or \ 4$$

It would therefore appear that it is relatively easy to move fairly large loads. This is not the

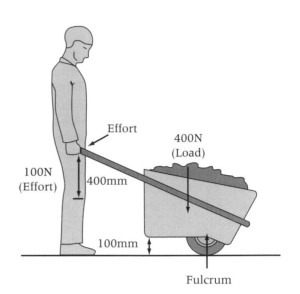

Fig. 3.20 A wheelbarrow

case, however, since you have to move the effort a greater distance than the load itself. When you are comparing these two distances, you are calculating the **velocity ratio (VR)**:

$$VR = \frac{distance \ moved \ by \ effort}{distance \ moved \ by \ load}$$

Using the wheelbarrow in Figure 3.20 again, the velocity ratio can be calculated:

$$VR = \frac{distance \ moved \ by \ effort}{distance \ moved \ by \ load} = \frac{400mm}{100mm} = 4:1 \ or \ 4$$

The overall **efficiency** of a mechanism is a comparison between what you put in and what you get out. It is calculated by:

$$Efficiency = \frac{MA}{VR} \times 100\%$$

In the wheelbarrow example, the efficiency is:

$$Efficiency = \frac{4 \ (MA)}{4 \ (VR)} \times 100\% = 100\%$$

In reality, however, no mechanism is 100 per cent efficient since levers bend and belts rub and twist.

Class 3 type levers are not used in situations where large loads are involved because they have a mechanical advantage of less than 1.

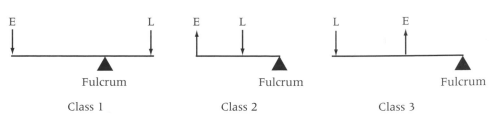

Figure 3.21 Schematic representation of classes of levers

This is because a greater effort is required than that of the load which is being moved.

Linkages

Linkages are forms of levers that are used to change the direction of motion within a system because the output movement is opposite to the input movement. Depending upon the position of the fulcrum or pivot point, the size of movement can be amplified. Linkages can also be used to transfer motion through 90 degrees. This type of linkage is known as a *bell crank*.

Toggle clamps are used to obtain large forces to lock things into place, such as the plastic sheet on vacuum forming machines. When a force is applied to the toggle, the mechanism is pushed out until the levers lock when they are in a straight line or just over centre.

EXAMINATION QUESTIONS

Example questions and answers.

Q1
a) *Give an example of a Class 2 type lever.*
b) *Briefly explain the principles of a Class 1 lever.* **(2 marks)**

Acceptable answer
a) A **wheelbarrow** is a Class 2 type lever.
b) For a **little effort, much larger loads can be moved**; however, **the effort must be moved much greater distances than the load**.

Gear systems

You need to

- [ ] **understand how gear systems can be used to change speed and direction of rotation**
- [ ] **be able to identify and describe simple and compound gear trains**
- [ ] **be able to select appropriate types of gears for specific applications**
- [ ] **calculate simple gear ratios and transmission speed.**

KEY TERMS
Check you understand this term

idler

Further information can be found in *Advanced Design and Technology for Edexcel, Product Design: Resistant Materials Technology*, Unit 3B3, section 4.

KEY POINTS

Gear systems and gear trains

A gear wheel is a basic mechanism that, when coupled together with other gear wheels, transmits rotary motion and force, and can change the direction of the motion. When two gear wheels are brought together, their teeth mesh. The two wheels rotate in opposite directions at speeds depending upon their sizes. If a driver gear with 50 teeth, rotating at 100rpm (revolutions per minute), meshes with a smaller gear of 25 teeth, the smaller driven gear will rotate at 200rpm in the opposite direction. This is known as the velocity ratio and can be calculated by the formula:

$$VR = \frac{\text{number of teeth on driven gear}}{\text{number of teeth on driver gear}} = \frac{25}{50} = \frac{1}{2}$$

If the driver and driven are both required to rotate in the same direction, then an **idler** gear is required. An idler gear has no impact on the speed of rotation of the system.

Types of gears

- A *simple gear train* consists of two gears meshing together. The input and output are in opposite directions and the velocity ratio is dependent upon the relative sizes of the two individual gears.
- A *compound gear train* consists of two separate gear trains meshed together. This invariably involves two gears mounted on the same shaft. In this arrangement, very large increases or decreases in speed can be achieved.

Table 3.29 Types of gears and applications

Gear type	Diagram	Applications
Spur gears/crown wheel		Used in simple motors/gearboxes for children's toys. Speeds can be increased/reduced and direction of rotation can be changed. Two spur gears together are termed a simple gear train. A crown wheel is another term for a single gear wheel.
Bevel		Bevel gears have their shafts at 90° to each other. This allows rotational direction/forces to be transmitted through 90°. If different sized gears are used, the speed of rotation can be increased or decreased. They are used in hand drills in school workshops.
Helical		These teeth are cut at a slight angle. They run more quietly and are generally more efficient; they are capable of transmitting larger forces. They are used in automotive gearboxes for large lorries.
Worm		Like bevel gears, worm gears are used to transmit force and motion through 90°. Large reductions/increases are achieved because the worm wheel is considered to have only one tooth. They are used in food mixers to turn the blending whisks.
Rack and pinion		A rack and pinion is used to change rotary motion into linear motion. They are used in canal locks to let water in and on pillar drills in order to move the rotating drill down into the workpiece.

Calculating simple gear ratios and transmission speeds

Transmission speeds can be calculated quite easily using the formula:

$$VR = \frac{\text{velocity of driven}}{\text{velocity of driver}} = \frac{\text{number of teeth on driven gear}}{\text{number of teeth on driver gear}}$$

EXAMINATION QUESTION

Example question and answer.

Q1 *A driver gear is rotating at 120rpm and has 60 teeth. It meshes with the driven gear, which has 40 teeth. Calculate the transmission speed or velocity of the driven gear.* **(2 marks)**

Acceptable answer

$$VR = \frac{\text{velocity of driven}}{\text{velocity of driver}} = \frac{\text{number of teeth on driven gear}}{\text{number of teeth on driver gear}}$$

$$\frac{\text{velocity of driven}}{120} = \frac{40}{60}$$

$$\text{velocity of driven} = \frac{40 \times 120}{60} = 80\text{rpm}$$

Resistors

You need to

- ☐ understand that fixed and variable resistors control current and voltage in circuits
- ☐ recognise that thermistors and light-dependent resistors are semi-conductors whose resistance changes with temperature and light levels
- ☐ know how to read the resistor colour code to determine resistor values.

KEY TERMS

Check you understand this term

resistors

Further information can be found in *Advanced Design and Technology for Edexcel, Product Design: Resistant Materials Technology*, Unit 3B3, section 5.

KEY POINTS

Fixed and variable resistors

Resistors are available in many forms. The two most common forms are variable and fixed. Resistors have two main functions:

- to control the amount of current flowing in a circuit
- to drop voltages in circuits.

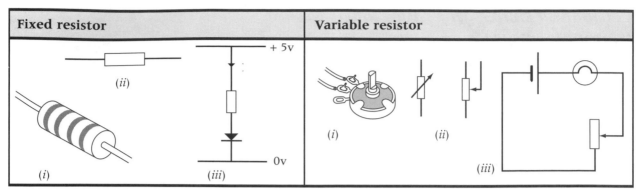

Fig. 3.22 Fixed and variable resistors

All types of resistors are rated according to three factors:

- a resistance value in ohms
- a tolerance
- a power rating in watts.

The amount of current (I) flowing through a resistor is governed by the voltage across its end and the resistor value. This can be expressed by the formula:

$$I = \frac{V}{R}$$

where

V is the voltage across the resistor
I is the current flowing through it
R is the resistance of the resistor.

It is normal practice to measure current in mA since we only deal with very small currents. Similarly, resistance is usually measured in kilohms (kΩ), where 1 kilohm is 1000 ohms:

1000 ohms $= 1k\Omega$
1/1000th amp $= 1mA$

A fixed resistor by definition has a single value, of 1 kΩ, for example. A variable resistor is capable of having any value of resistance between 0Ω and 1kΩ, for example (Figure 3.22).

Thermistors and light-dependent resistors

Thermistors and light-dependent resistors (LDRs) are two other forms of resistors. Since they are made from semi-conductive materials,

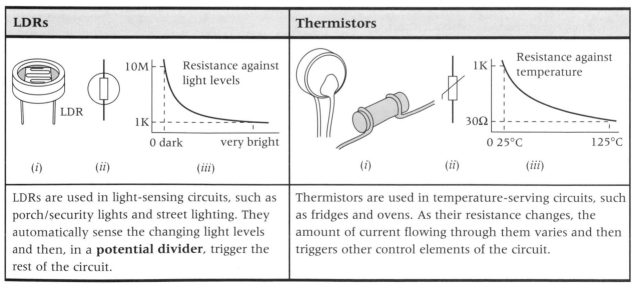

LDRs are used in light-sensing circuits, such as porch/security lights and street lighting. They automatically sense the changing light levels and then, in a **potential divider**, trigger the rest of the circuit.

Thermistors are used in temperature-serving circuits, such as fridges and ovens. As their resistance changes, the amount of current flowing through them varies and then triggers other control elements of the circuit.

Fig. 3.23 LDRs and thermistors

their resistance changes in response to changing light levels and heat.

Resistor colour codes

Resistor values range from 10Ω up to 10 million Ω. Their values can be identified by the series of colour bands that run around their bodies. The most common type of resistor has four coloured rings.

Band 1 and Band 2 will give a number between 0 and 99, which is then multiplied by the value in Band 3. Since the value of the resistor is never exactly the value indicated by the first three bands, the fourth band indicates a tolerance level, which the value will fall between.

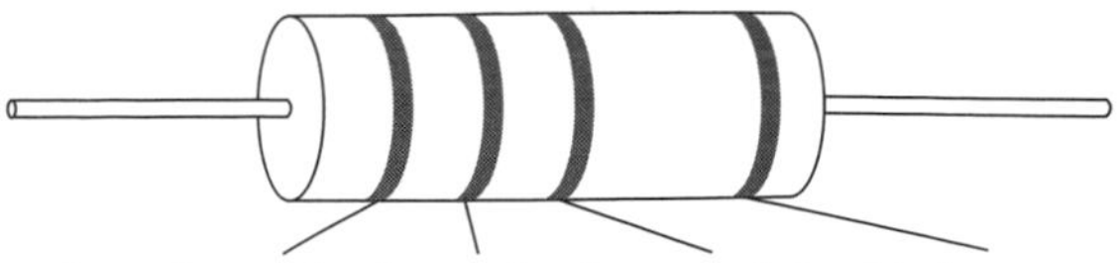

Colour	Band 1	Band 2	Band 3	Tolerance
black	0	0	x1	
brown	1	1	x10	
red	2	2	x100	
orange	3	3	x1000	
yellow	4	4	x10,000	
green	5	5	x100.000	
blue	6	6	x1,000,000	
violet	7	7		
grey	8	8		
white	9	9		
silver				±10%
gold				±5%

Resistor colour code

Fig. 3.24 Resistor colour codes

EXAMINATION QUESTION

Example question and answer.

Q1 *A resistor is coloured brown, black, red and silver. Work out the value of the resistor including its upper and lower limit.* **(3 marks)**

Acceptable answer
1000Ω or 1kΩ
Upper limit 1000 + 100 = 1100Ω or 1.1kΩ
Lower limit 1000 − 100 = 900Ω

Potential dividers

You need to:

☐ be able to use resistors in series to create a potential divider
☐ be able to use sensors and a variable resistor in series to create an adjustable sensing input.

KEY TERMS
Check you understand these terms

series, adjustable sensing inputs

Further information can be found in *Advanced Design and Technology for Edexcel, Product Design: Resistant Materials Technology*, Unit 3B3, section 6.

KEY POINTS

Resistors in series

If two resistors are connected in **series**, that is, one after the other, they can be used to divide the potential difference across them.

The value of V_{out} in the potential divider shown in Figure 3.25a will depend upon the ratio of R1 to R2 and the voltage supply. Obviously, if R1 = R2 and the voltage supply is 10V as shown, V_{out} will be 5V.

If, however, the values of R1 and R2 are not equal, V_{out} will be somewhere between the supply voltage and 0V. If, using the potential divider shown in Figure 3.25a, R1 = 330Ω and R2 = 680Ω, then V_{out} can be calculated in the following way.

Using I = V/R, the current can be calculated

I = 10/1010 = 9.9mA

V_{out} can now be calculated by I × R2:

9.9mA × 680Ω = 6.73V

An alternative method is to use the formula

$$V_{out} = V_s \times \frac{R2}{(R1 + R2)}$$

where

V_s = supply voltage

Adjustable sensing inputs

Adjustable sensing inputs are another form of potential divider made from LDRs and thermistors with a variable resistor, as shown in Figure 3.25b and c. As the light and temperature levels fluctuate, the resistance levels change, therefore affecting the total resistance value. This in turn will alter V_{out} and this change can be used to trigger a transistor as either a switch or amplifier.

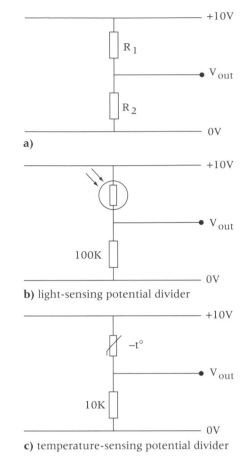

a)

b) light-sensing potential divider

c) temperature-sensing potential divider

Fig. 3.25 Adjustable sensing inputs

EXAMINATION QUESTION

Example question and answer.

Q1 *Using Figure 3.25b, calculate the value of V_{out} if the LDR has a resistance of 50kΩ.* **(2 marks)**

Acceptable answer

$$V_{out} = V_s \times \frac{R2}{(R1 + R2)}$$ where V_s = 10V, R1 = 50kΩ and R2 = 100kΩ

$$V_{out} = 10 \times \frac{100K}{(50K + 100K)} = 6.66V$$

Diodes

You need to

☐ **know that diodes allow current to flow in one direction only**
☐ **understand that diodes are used in rectification and to protect devices from back electromotive force (emf)**
☐ **understand the action of an LED.**

KEY TERMS

Check you understand this term

back electromotive force (emf)

Further information can be found in *Advanced Design and Technology for Edexcel, Product Design: Resistant Materials Technology*, Unit 3B3, section 7.

KEY POINTS

Current flow in diodes

Diodes are semi-conductors that only allow current to flow through them in one direction (Figure 3.26a). Since they only have two legs, it is essential that they are inserted into circuits the correct way around.

Diodes as rectifiers and protective devices

When an alternating current (AC) is converted into a direct current (DC), rectification and smoothing occur. There are two types of rectification: *half wave* and *full wave*.

A *half wave rectifier* (Figure 3.26b) is essentially a voltage divider that contains a diode and a resistor. When the input wave is positive, the diode is forward biased, there is very little resistance and therefore V_{out} is virtually the same as V_{in}. When the input is negative, the diode is reverse biased, a large resistance prevails and V_{out} becomes 0V. The output formation is therefore half the input, hence the term, half wave rectifier.

A *full wave rectifier* (Figure 3.26c) involves four diodes. When forward biased, the current follows the path A, D_2, R_L, D_3, B and when negative, B, D_4, R_L, D_1, A, which results in the output being a fully rectified wave formation. Rather than using four separate diodes for full wave rectifiers, a small package known as a bridge rectifier is available.

Diodes are also used to protect transistors when they are used to switch motors and relays (Figure 3.26d). When either a relay or motor is turned off, a large voltage known as a **back electromotive force (emf)** is produced. This induces a voltage in the coil, causing a current to flow backwards, which can cause damage to the transistor. To overcome this, a diode is placed across the relay.

Light-emitting diodes (LEDs)

An LED is a light-emitting diode that emits light when a current passes through it. LEDs require

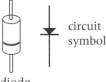

a) diode

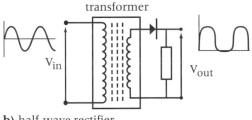

b) half wave rectifier

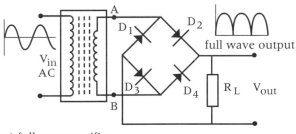

c) full wave rectifier

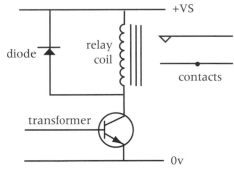

d) Diode used to protect a transistor

Fig. 3.26 Diodes and their applications

very little current in order to make them work and last much longer than conventional bulbs. However, they must always be used in series with a current-limiting resistor. The LED needs 2V across it to make it work. However, if a reverse bias of more than 5V is applied across an LED or the current exceeds 50mA, it will blow.

EXAMINATION QUESTION

Example question and answer.

Q1 *An LED is connected to a 10V supply. The LED is rated at 10mA and 2V. Calculate a suitable value for a protective resistor.* **(2 marks)**

Acceptable answer

Using V = IR: R = V/I
Since the LED requires 2V, the resistor needs to drop 8V at 10mA:
R = 8V/10mA = 800Ω

Capacitors

You need to

☐ **know that capacitors store an electrical charge**
☐ **understand how polarised capacitors are identified and used in circuits**
☐ **know that resistor/capacitor networks are used in time delay circuits.**

KEY TERMS

Check you
understand these terms

time delay networks, time constant

Further information can be found in *Advanced Design and Technology for Edexcel, Product Design: Resistant Materials Technology*, **Unit 3B3, section 8.**

KEY POINTS

Capacitors storing electrical charge

A capacitor can be used to store an electric charge. The larger the capacitor, the larger the charge it can store. The charge-storing ability is known as *capacitance* (C) and is measured in farads (F). However, one farad is a large unit and it is much more common to use microfarads (μF), where $1\mu F = 1 \times 10^{-6}$ or 1/1000 000F.

Once a capacitor has been charged by connecting it to a battery, it will retain the voltage until it is discharged or leaks. The amount of charge a capacitor can store, C, is calculated by:

C = Q/V

where

C = capacitance in farads (F)
Q = charge in coulombs (C)
V = voltage in volts (V).

Capacitors are categorised by two factors, which are printed on the side of the capacitor itself: their capacitance and their working voltage.

Polarised capacitors in circuits

Capacitors are made from many different types of materials and exist in many forms. Polarised capacitors (Figure 3.27a) are among the most common. Since they are polarised, they must be handled and inserted into circuits the correct way around. To help identify which way to insert them, the cathode is the shorter one and a −ve symbol is printed alongside it.

Resistor/capacitor networks in time delay circuits

Capacitors and resistors are connected in series in circuits to form **time delay networks** (Figure 3.27b). As the capacitor charges, time passes and the delay is created. The time taken for the capacitor to charge to two-thirds of the supply voltage is called the **time constant** (Figure 3.27c). This can be calculated by:

a) electrolytic capacitor

circuit symbol

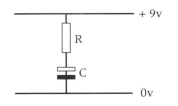

b) RC network

Fig. 3.27 Capacitors

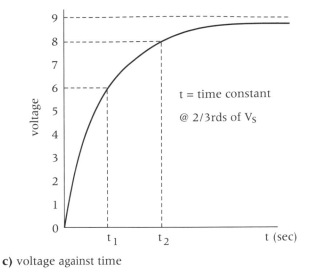

c) voltage against time

t = CR

where

t is time in seconds (s)
C is capacitance in farads (F)
R is resistance in ohms (Ω).

In the combination in Figure 3.27c, it takes approximately five time constants for the capacitor to reach the supply voltage. If the resistor were to be replaced with a variable resistor, the time delay could be varied and made adjustable.

EXAMINATION QUESTION

Example question and answer.

 *A 100kΩ resistor and a 10µF capacitor are connected in series. Calculate the time constant for this series network.* **(2 marks)**

Acceptable answer
T = R × C = 100kΩ × 100µF = 10 seconds

Transistors

You need to

☐ understand how to use an NPN transistor as an electronic switch and current amplifier.

 KEY TERMS
Check you understand this term

threshold voltage

 **Further information can be found in** *Advanced Design and Technology for Edexcel, Product Design: Resistant Materials Technology,* **Unit 3B3, section 9.**

KEY POINTS

NPN transistors as electronic switches and current amplifiers

Transistors are semi-conductors and therefore behave both as insulators and conductors. A

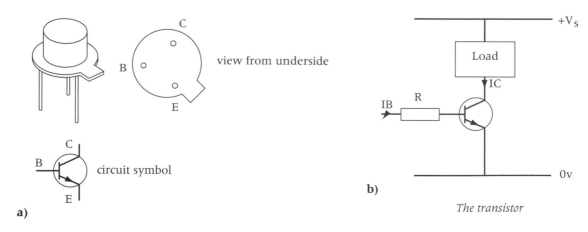

Fig. 3.28 Transistors

transistor is a three-legged component held in a metal can. The three legs can be identified by viewing the transistor from underneath. The pin that is nearest the tab is the emitter, followed by the base and then the collector (E, B and C in Figure 3.28a).

It is essential that each leg of the transistor is connected to the correct part of the circuit. The base of the transistor must always be protected by a current-limiting resistor since too much current will blow the transistor. The base is the pin that detects the charges from the potential divide, such as those created with LDRs and thermistors. When the voltage on the base rises above 0.6V, the transistor switches on; this is known as the **threshold voltage**. At this point, the transistor behaves as a conductor rather than an insulator.

Transistors can also be used as current amplifiers. The gain of the transistor is a measure of its amplifying capacity. This is the ratio of the current flowing into the transistor, I_B, to that which flows in, I_C, as shown in Figure 3.28b. This varies from type to type. The gain of a BC108 is between 200 and 800, which is not very precise, so that I_C could be up to 800 times bigger than I_B. The gain of a transistor can be calculated using:

$$h_{fe} = I_C/I_B$$

Transistors are chosen by either their gain or for the amount of current that can pass through, I_C, without burning it out. It is not possible for a transistor to have both a high gain and a high current handling capacity.

EXAMINATION QUESTION

Example question and answer.

Q1 *Give the **two** primary functions of a transistor.* **(2 marks)**

Acceptable answer
A transistor is **used as a current amplifier** and an **electronic switch**.

Relays

You need to

☐ **know that relays are used to interface primary and secondary circuits using mechanical switch contacts.**

KEY TERMS
Check you understand this term

interface

📖 **Further information can be found in *Advanced Design and Technology for Edexcel, Product Design: Resistant Materials Technology*, Unit 3B3, section 10.**

KEY POINTS

Relays to interface primary and secondary circuits

A relay contains an electromagnet that is activated by passing a current through the relay. The primary purpose of a relay is to act as an **interface** device between a primary and secondary circuit.

A transistor is used as an electronic switch to activate the relay. When the coil of the relay is energised, the contacts are closed by a mechanical switching action and the secondary part of the circuit is completed.

In general, the secondary part of the circuit often operates at a higher voltage than the primary part and therefore has to be kept separated.

EXAMINATION QUESTION

Example question and answer.

Q1 *The circuit diagram shown below uses a sensor circuit to control a relay. The relay operates a mains-operated electric fan. Explain why the relay is used in this circuit.* **(2 marks)**

Acceptable answer

Because the circuit operates on **two separate voltages, 9V on the primary side and mains voltage on the secondary side**, the relay is used as an **interface device between the two sides**.

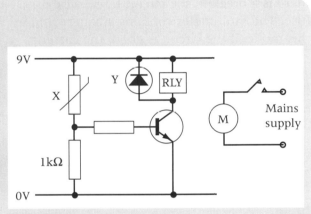

Fig. 3.29 A relay circuit

'London Qualifications Ltd' accepts no responsibility whatsoever for the accuracy or method of working in the answer given.

Circuit testing

You need to

☐ be able to use a multimeter to test for continuity and resistance, voltage levels and current flow in a circuit.

KEY TERMS
Check you understand this term

continuity

📖 **Further information can be found in *Advanced Design and Technology for Edexcel, Product Design: Resistant Materials Technology*, Unit 3B3, section 11.**

KEY POINTS

Use of a multimeter for testing a circuit

A multimeter has many functions. It can be configured to test and measure continuity, resistance, voltage and current.

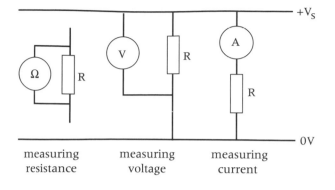

measuring   measuring   measuring
resistance   voltage   current

Fig. 3.30 The correct use of a multimeter

A **continuity** test will simply indicate whether there is a break in the circuit track, wire or fuse. The multimeter contains a battery and in this mode, if the circuit is good, the current will flow because the circuit is complete. A small audible sound can be heard to indicate that all is well. If no sound is heard when testing a fuse, this indicates that the fuse has blown.

When measuring a component's resistance, it is essential that the component is removed from the circuit, otherwise the circuit will affect the resistance. The component to be measured should simply be held against the two meter probes and the meter set to read resistance.

The potential difference across a circuit or two points of a circuit is measured in volts. For both AC and DC, the meter should be connected in parallel with the two points. This will ensure that it is measuring the voltage between the two points being tested.

Since current is the movement of electrons through a conductor, when measuring current, the meter must connect in series to detect the flow. A physical break must be made in the circuit and the meter used to join the circuit up again so that the current has to flow through the meter.

EXAMINATION QUESTION

Example question and answer.

 *When measuring resistance, the component must be taken out of the circuit. Explain **one** reason for this.* **(2 marks)**

Acceptable answer
If the component is not removed from the circuit, **the reading will be affected by other components on the circuit board** and this will **influence the overall resistance reading**.

PRACTICE EXAMINATION STYLE QUESTION

1 The diagram below shows a microswitch. An **effort** force must be applied to operate it.

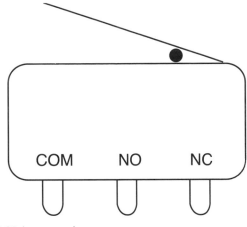

Fig. 3.31 A microswitch

a) Copy the diagram, adding labels to show the **effort**, the **load**, and the **fulcrum**. **(3)**

b) State the class of lever involved in this device. **(1)**

c) Give **one** other example of this type of lever. **(1)**

Figure 3.32 shows a lever in action.

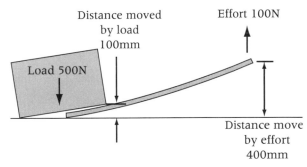

Fig. 3.32 A lever

d) Calculate, using the information given:
   i)   the mechanical advantage
   ii) the velocity ratio
   iii) the efficiency of the system. **(5)**

e) Describe **two** ways power may be generated using water. Using a schematic diagram, show in detail how **one** of your chosen methods works. **(5)**

2 Gears are toothed wheels, fixed to the driver and driven shafts, which mesh together. A simple gear train is shown in Figure 3.33. The driven gear has 24 teeth and the driver gear 12.

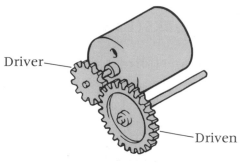

Driver

Driven

Fig. 3.33 Simple gear train

a) i) Calculate the velocity ratio of the gear train. **(3)**

ii) Using notes and diagrams, explain how the driven and driver gear can be made to rotate in the same direction. **(3)**

iii) Describe **two** different types of gears that can be used to change rotary motion through 90°. Give an application for each of the two methods given and a simple schematic sketch. **(6)**

b) A different type of gear system is shown in Figure 3.34. Name the gear system shown and give an example of an application. **(3)**

Fig. 3.34

Total for this question paper: 30 marks

Part 3
Advanced GCE (A2)

UNIT 4a

Materials, components and systems (R401)

Unit 4 is divided into two sections:

- Section A: Materials, components and systems (compulsory for all candidates)
- Section B: Consists of three options (of which you will study only one).

Section A will be assessed during the $1\frac{1}{2}$ hour, Unit 4 examination. You should spend half of your time answering all of the questions for this section. It is important to use appropriate specialist and technical language in the exam, along with accurate spelling, punctuation and grammar. Where appropriate, you should include clear, annotated sketches to explain your answer.

Selection of materials

You need to

know about the relationship between the characteristics, properties and materials choice relating to:

- ☐ quality
- ☐ manufacturing processes
- ☐ material limitations
- ☐ wear and deterioration
- ☐ maintenance
- ☐ life costs.

KEY TERMS

Check you understand these terms

quality, wear, deterioration, maintenance, life costs

Further information can be found in *Advanced Design and Technology for Edexcel, Product Design: Resistant Materials Technology*, Unit 4A, section 1.

KEY POINTS

Quality

When choosing and selecting materials for products, full consideration must be given to the relationship between the working characteristics and the properties of the materials. **Quality** is a key aspect in the selection process and relates to the overall quality of the material to be used. For example, a piece of cupped and warped oak would not be appropriate for use in a table top. Quality of performance would be measured against the material's performance in tests that take account of **wear** and **deterioration** over a period of time, such as nylon drawer runners or bearings. The manufacturing processes that materials are subjected to will have a significant impact on the overall quality of the product and its fitness for purpose. For example, sand casting is not an appropriate process for the manufacture of a pencil sharpener because of the product's intricate shape and size and the required level of surface finish and dimensional accuracy.

Manufacturing processes

The manufacturing processes available for the production of components and products are significantly influenced by the initial choice of material. On occasions, if the material has already been decided upon, it may indeed limit the manufacturing processes.

The scale of production will also influence the range of processes that can be selected. Table 4.1 gives an indication of the production level of a range of manufacturing processes.

Material limitations

Materials are tested to physical destruction in order to establish the limits to which they can be loaded, stretched and compressed. A chair must withstand the weight of someone sitting on it, and even swinging on it or putting their weight only on two chair legs. If the chosen material was not capable of withstanding these forces, the chair would fail as a structure and would collapse when somebody sat on it or leaned back. One way of compensating for people abusing products such as chairs is to build in a factor of safety. Thus a chair is made four or six times stronger than it actually needs to be, which is why products appear to be over-engineered. When casting, injection moulding and blow moulding, flow rates of the material into the mould or cavity are critical.

Wear and deterioration

Materials wear out and deteriorate. Wear is particularly likely where surfaces rub and come into contact with other surfaces. Bearings and bearing surfaces are prone to wear, leading to further mechanical breakdown and malfunction. When the clutch in a car wears, the driver cannot change gear and there is an unpleasant burning smell. Tools with sharp edges wear, but can often be re-sharpened a number of times before they can no longer be used. All materials to some extent deteriorate – the colours may fade or the material may rust, rot and finally, in the case of timbers, decompose.

Wear and deterioration can be reduced by increasing the hardness of surfaces that are in contact, and also by protecting the material surface from the natural elements, which accelerate the deterioration process.

Maintenance

The **maintenance** of materials and products must be given consideration during the initial selection process. The material's surface finish will protect it to some extent, but most surface finishes, paint especially, must be carefully maintained in order to preserve and extend the product's useful working life. Finishes vary in their resistance to deterioration; some surface treatments, such as galvanising, can break down if they become chipped or damaged.

Similar products, such as garden seating, can be made in wood, plastic or metal. All have their merits but each has very different maintenance requirements. It is also essential that products can be serviced and maintained throughout their working life. Worn-out parts should be replaceable, and parts such as blades able to be sharpened or replaced.

Table 4.1 Manufacturing processes related to level of production

One-off	Batch	High volume
• Sand casting • Carving • CNC manufacture • Spark erosion • Laminating • Forging	• Sand casting • Die casting • CNC manufacture • Spark erosion • Laminating • Sintering • Stamping • Forging	• Die casting • CNC manufacture • Injection moulding • Blow moulding • Sintering • Stamping

Life costs

Life costs take into account many moral and environmental issues. The damage to the environment caused by harvesting raw materials, such as timber, oil, and ores for use in metal production, should be kept to an absolute minimum. Handling of the product at the end of its useful working life must also be considered. Government and European legislation now forces companies and manufacturers to take more responsibility for the recycling of their products and to design products that can be reused and recycled.

Sustainability is also an issue. In particular, timbers should only be used from managed sources. Minimising levels of pollution in the production of materials, such as plastic and metals, must also be considered when selecting a material.

EXAMINATION QUESTION

Example question and answer.

Q1 *Identify the factors a designer might take into account when selecting the materials for a garden bench.* **(5 marks)**

Acceptable answer

Any material chosen for the bench must be **hard-wearing and long-lasting**. The **surface treatment** and finish must also be considered since all coatings on woods and metals must be maintained to protect the material from natural elements such as rain and sunshine. This will **affect the overall integrity and strength of the material and would tend to favour the use of plastics**.

The **scale of production** will be a major factor in the choice of material. Some materials and processes are more appropriate to high volume production (for example, injection moulding). Others are more suitable for batch and one-off (for example, forging, scrollwork and traditional woodworking joints).

The **sustainability of the material is also important**. Wood should be used from **managed forests, and recycled metal and plastic should be used wherever possible**. **Disposal of the materials** used and their **potential to be recycled should be given full consideration** as these matters are governed by European legislation.

New technologies and the creation of new materials

You need to

know about modern and smart materials, such as:

☐ **tinted glass and reactive glass**
☐ **solar panels**
☐ **thermo-ceramics**
☐ **shape memory alloys**
☐ **LCD displays**
☐ **piezo-electric actuators**
☐ **smart composite materials**
☐ **composite materials**
☐ **new materials used in the computer and electronics industry.**

KEY TERMS
Check you understand these terms

photochromic glass, photovoltaic cell, thermo-ceramics, shape memory alloys, liquid crystal displays, smart materials, composite materials, silicon

Further information can be found in *Advanced Design and Technology for Edexcel, Product Design: Resistant Materials Technology*, Unit 4A, section 2.

KEY POINTS

Tinted glass and reactive glass

Tinted glass contains additives that make it slightly darker than conventional clear glass. **Photochromic glass** is clear but darkens in response to exposure to ultraviolet light. This type of glass is used in spectacles, which react quite slowly.

Solar panels

It is estimated that the sun generates over 2000 times as much energy as the current world demand. This solar energy can be harnessed by using solar panels or **photovoltaic cells** (Table 4.2).

Thermo-ceramics

Thermo-ceramics are ceramics that can withstand very high temperatures. Silicon nitride is a thermo-ceramic used for turbine blades and turbo-charge units in the automotive industry. Like most ceramics, it is very brittle. Research into its use for automotive cylinder blocks is underway.

Shape memory alloys

Shape memory alloys were discovered in the late 1970s. As a material, they can be plastically deformed, but they return to their original pre-deformed shape on reheating above a certain temperature, which can be as low as 70°C. This process can be repeated indefinitely; the heat required is usually produced by passing a small current through the material, which is often a wire. Nitinol (nickel-titanium) contracts by

approximately five per cent of its length when heated. These materials are used to seal hydraulic tubes onto pipes and as valve controls in hot water mixing systems.

LCD displays

Liquid crystal displays (LCDs) are used extensively as numerical and alphanumerical displays and indicators. Liquid crystals are organic, carbon-based compounds that exhibit liquid or solid characteristics depending upon whether a voltage is being applied. As a method of display, they require much smaller currents than conventional seven-segment LED (light emitting diode) displays.

The numbers or letters appear on the display by applying a voltage to certain segments, which darken in relation to the silver background. LCD displays are used in calculators and for mobile phone screens.

Fig. 4.1 LCD screen on a mobile telephone

Table 4.2 Methods of harnessing solar energy

Solar panels (water heating)	Photovoltaic cells
• Cheaper to set up and install • Painted black to absorb heat • Water pumped around system in thin tubes gets hot • Water fed back into the domestic hot water system	• Initially expensive to purchase • Generates electrical current, or used to charge batteries, or fed into national grid • Used in space to power satellites • 'Free' energy once cells purchased • Uses silicon, which reacts when exposed to intense light • Large areas needed to be of real use

Piezo-electric actuators

Piezo-electric actuators work in two different ways: they can produce a voltage when tapped or pressed, or produce movement in response to an applied voltage. Uses include producing the sound in musical greetings cards and as input transducers in burglar alarms, activating the alarm when trodden on or disturbed.

Smart composite materials

'**Smart materials'** is a term covering a wide range of materials whose physical properties can be varied by an input. These include piezo-electronic actuators and shape memory alloys.

Composite materials

Carbon fibres mixed with a resin and glass reinforced plastics (GRPs) are both examples of **composite materials**, formed when two or more materials are combined by bonding.

GRP is available in several forms: a loosely woven fabric, a string of filaments wound together, a matting of short fibres, or loose, short, chopped strands. Usually, the glass fibre strands are placed over or inside a former or mould. They are subsequently covered in a polyester resin containing a hardener element and other additives such as pigments. As the layers are built up and soaked in resin, they are squeezed together. This manufacturing process is ideal for large structures, such as boat hulls,

giving good strength and rigidity and low weight.

Composites based on carbon fibres have become widely used in the Formula One racing industry. Their immense strength and very low weight make them ideal for use in racing cars as structural supports and ties. Their very high quality of surface finish offers little resistance to airflow. The process is similar to GRP moulding, although the raw forms are very different: glass fibre strands are randomly orientated, while carbon fibres are neatly ordered.

New materials used in the computer and electronics industry

About 28 per cent of the Earth's crust consists of **silicon**, in the form of silicon oxides, such as sand, quartz or rock. Silicon is used as an alloying element in the production of steel, but its main use is in the electronics industry. It is a semi-conductor, which can behave as either an insulator or as a conductor. It is therefore used in the manufacture of silicon chips and other components ,such as transistors and diodes.

Silicon has been central to the development of the speed and processing power of computers. Computers have their entire central processing unit (CPU) made from a single integrated circuit containing more than a million transistors. The same computers also need memory chips of ever-increasing capacity.

EXAMINATION QUESTIONS

Try answering these yourself.

*Give **one** advantage of using an LCD rather than a seven-segment light emitting diode display.* **(2 marks)**

Describe the difference between carbon fibre and glass fibre matting with reference to how the fibres are aligned and orientated. **(2 marks)**

Impact of modern technology and biotechnology

You need to

know about:

☐ genetic engineering in relation to woods

☐ the use of micro-organisms to aid the disposal of environmentally friendly plastics

☐ the recycling of materials.

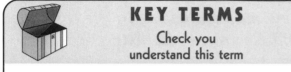

KEY TERMS

Check you understand this term

biotechnology

Further information can be found in *Advanced Design and Technology for Edexcel, Product Design: Resistant Materials Technology*, Unit 4A, section 2.

KEY POINTS

Genetic engineering in relation to woods

Biotechnology has enabled scientists to investigate the possibilities of genetically engineering and modifying timbers. In the future, faster growing trees may increase timber supplies and genetically modified trees could produce timber more resistant to wear, rot and animal infestation. This will be of particular benefit in the construction industry where the majority of timber used has to be treated with preservatives prior to use. However, this would bring problems, such as the disposal of a material that will not rot.

Paper production uses an enormous amount of timber. The removal of lignin, the natural substance that holds the fibres together, with highly toxic chemicals is costly and damaging to the environment. Reducing genetically the amount of lignin produced would be advantageous.

The use of micro-organisms to aid the disposal of environmentally friendly plastics

One major disadvantage of plastics is disposal. Recent developments in biotechnology have brought a new generation of plastics. Although some biodegradable plastics have been on the market for more than 60 years, there has been increasing pressure to develop plastics from materials other than petrochemicals and to look at their subsequent disposal.

Many new-generation plastics are manufactured from natural hydrocarbons, so they can easily be broken down by micro-organisms that can be found in compost heaps and on open cultivated land. Farmers now use plastic film to warm and protect crops in winter and early spring. By the time the crops are harvested, the plastic sheeting has started to decompose. It is eventually ploughed back into the ground and totally decomposes.

The recycling of materials

Recycling is a major issue facing designers and manufacturers today. Life-cycle analysis now has to be carefully considered, right through to the disposal and recycling of the product. UK and European legislation now ensures that manufacturers have to carefully dispose of and recycle a certain percentage of any product, from washing machines to cars, once it reaches the end of its useful working life.

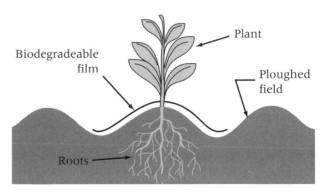

Fig. 4.2 Use of biodegradable film in the farming industry

Modification of properties of materials

You need to

☐ **know about modification of properties of materials.**

KEY TERMS

Check you understand these terms

co-polymerisation, foamants

📖 **Further information can be found** in *Advanced Design and Technology for Edexcel, Product Design: Resistant Materials Technology*, **Unit 4A, section 2.**

KEY POINT

Modification of properties of materials

Metals

Metals can be modified in several ways: alloying, heat treatment, work hardening, age hardening and sintering. An alloy is a combination of one metal with one or more other metals and, in some instances, non-metals. Alloys enhance or improve the properties of the initial 'parent' metal, such as hardness, toughness or improved resistance to decay. There are many forms of heat treatment, each serving a particular purpose (Table 4.3).

Work hardening

Work hardening is the hardening of a piece of metal by the very process of working on it.

Table 4.3 The heat treatment of steels

Hardening	Increases the material's resistance to indentation and abrasion. The steel, which must contain more than 0.4% carbon, is heated to a temperature above its upper critical temperature before being quenched in water, oil or brine. Now very hard, it is also very brittle.
Tempering	The hardened material is cleaned and the surface polished. It is heated to a particular temperature, depending upon what it is going to be used for, and then quenched in water, oil or brine.
Annealing	As the metal's grain structure is deformed by cold working, the only way to restore the original structure is to anneal it. The piece is heated and held at that temperature before being allowed to cool in the open air.
Normalising	After the metal has been forged or work hardened, the grain structure must be normalised. The piece is heated, soaked at this temperature and then cooled in the oven.

This might be by pressing, punching or forging. In some instances, it is a useful side-effect since when car body panels are pressed, they are plastically deformed, and the material is also work hardened and its toughness is thereby increased. Hardening can also be carried out by:

- case hardening – a layer of carbon is fused into the surface of the component
- induction hardening – an induction coil is used to heat a localised area of the workpiece, often as it rotates, followed by a water jet
- flame hardening – similar to induction hardening, but a flame is used instead of an induction coil, followed by a water jet.

Age hardening

Age hardening is almost exclusively used for aluminium and its alloys. Despite what the term might actually suggest, there does need to be some heat involved in the process. After this initial heating and cooling period, the alloys will then harden with time.

Sintering

Sintering is widely associated with powder metallurgy. Fine powders are mixed to form alloys before they are compacted and sintered using heat and high pressure. Graphite is sometimes included as a lubricant since sintered components are often used as the bushes and bearings in vacuum cleaners, drills and other small mechanical products. One major advantage of sintered bearings is that they are considered to be self-lubricating because they retain oil.

Plastics

Co-polymerisation is when two or more monomers are combined to form a new material – a co-polymer. It has better properties than its component materials. Polyvinyl chloride (PVC) is a co-polymer, being formed from a mixture of vinyl chloride and vinyl acetate.

Cross linking is another method of increasing strength in plastics, for example, in the formation of PVC, where chlorine atoms have replaced the hydrogen atoms, improving the material's overall strength.

Plasticisers are added during the manufacturing of the plastic material to improve its ability to flow into the mould during injection moulding. They also reduce the softening temperature and as a general rule make the material less brittle. Fillers, fibres and **foamants** are used in plastics to increase properties such as strength and toughness and to reduce their density, making them ideal for use as buoyancy aids. Major disadvantages of early plastics were their deterioration and discoloration when exposed to ultraviolet light. This has been partly overcome by the addition of stabilisers, but over long periods they still discolour and fade.

Wood

Strength can be achieved in timbers by removing excess moisture once the tree has been felled. This process is known as *seasoning*. Seasoning can be accomplished either naturally in air or in a kiln (Table 4.4).

Laminating timber is another way of increasing the material's strength. Veneers or laminates are bonded together at right angles to form plywood, which is a very strong material in all lateral dimensions. Veneers and laminates can also be bonded together over formers to produce curved shapes, such as chair legs.

Table 4.4 Seasoning

Air seasoning	Kiln seasoning
• Slow and inaccurate • 25 mm of thickness takes one year • Moisture content only reduced to surrounding conditions • Does not kill all bugs • Can be infected by fungus • Takes up lots of space for a long time	• More expensive to kiln season • Carefully controlled moisture content • 25 mm of thickness takes only two weeks • Less space, greater throughput • Kills bugs and fungus

EXAMINATION QUESTION

Example question and answer.

Q1 *The properties of metals can be modified in different ways, such as alloying and hardening. Describe the **two** processes mentioned.* **(4 marks)**

Acceptable answer

Alloying – alloying is a process of **combining two or more metals in order to make a new material with properties that differ or are enhanced in comparison with the original materials.** An example of this would be **stainless steel,** **which is an alloy of steel and chromium.** Whereas steel rusts, stainless steel is resistant to this form of decay.

Hardening – this is **a process that physically increases the hardness of the material.** Steel has to **contain above 0.4 per cent carbon in order to be easily hardened.** This process involves **heating the steel to a temperature above its upper critical temperature,** often a glowing orange colour, **before being plunged into water, oil or brine. The higher the carbon content, the harder the steel will be.**

Values issues

You need to

☐ **understand the impact of values issues**
☐ **know the responsibilities of developed countries.**

KEY TERMS
Check you
understand these terms

developed countries, economics

📖 **Further information can be found in** *Advanced Design and Technology for Edexcel, Product Design: Resistant Materials Technology,* **Unit 4A, section 3.**

KEY POINTS

Impact of values issues

Every decision a designer or manufacturer takes when developing a new product will fall into one of the following categories:

- technical
- economic
- aesthetic
- social
- environmental
- moral.

Ultimately, the decision on whether to proceed with the development and subsequent manufacture will be assessed against the product's commercial viability – will it sell and make money?

Every need is a value judgement based on the criteria above, but for some companies the bottom line is profit. Some companies manufacture in countries where labour is extremely cheap in comparison to the UK, and many argue that this is morally wrong, especially when the products made are sold in **developed countries** at high or inflated prices. Is this correct? Other companies use materials from rainforests, leading to the destruction of natural habitats and to the loss of rare species. Animals are hunted and killed for their skins and horns, which are then used in exotic products sold for huge amounts of money. Is this acceptable?

Products for the disabled are highly specialised and in some instances have very small market demand. Despite being technically achievable, many products do not reach the market place because of **economics**: they will not make much money. Is this morally right?

Responsibilities of developed countries

Whether a country is considered 'developed' depends on whether it meets a certain number of criteria based on socio-economic factors. These measures generally relate directly or indirectly to technology, and, until recently, to their manufacturing technologies. This suggests that if a country is not 'developed', it hopes one day to become so. This status must be considered in relation to the impact on the environment and pollution it may bring. Legislation has been passed to ensure that all countries carefully monitor and control their environmental pollution, and careful consideration must be given to what happens to the product once it is no longer needed.

Wherever design and manufacture occur, all parties should take into account the need for:

- more effective and efficient use of materials
- greater re-use and recycling of waste products and materials
- reduced impact on environment
- safer disposal of waste and waste products
- greater efficiency and use of energy and natural energy
- greater use of sustainable resources, such as timber, from managed forests.

In essence, a balance has to be drawn between cost and benefit. The cost is not only monetary but also to the environment and global pollution.

EXAMINATION QUESTION

Example question and answer.

Q1 *Developed countries have a responsibility to consider the environment. Describe **one** environmental factor that should be taken into account when setting up major new facilities.* **(2 marks)**

Acceptable answer
Consideration must be given to the **local environment in respect of the noise, disruption and damage** the new facilities may bring and cause. **Increased traffic pollution** and **noise** will upset local people and **industrial pollution, such as fumes and waste water, may also cause disruption to the local eco-systems.**

PRACTICE EXAMINATION STYLE QUESTION

1 Give **two** different uses for silicon within the electronics and computer industries. **(2)**

2 Explain **two** advantages to the environment of using genetically modified timbers. **(4)**

3 a) Give **two** applications of carbon fibres. **(2)**

  b) Using **one** of the applications from your answer in (a), explain **two** reasons that make it appropriate for your named application. **(4)**

4 The properties of plastics can be modified by the following processes:
  * co-polymerisation
  * cross linking
  * plasticisers
  * fillers, fibres and foamants
  * stabilisers.
  Describe **two** of the above processes. **(4)**

5 New technologies are becoming increasingly important in all types of industry. Describe and give an example of the use of **two** of the following:
  * piezo-electric actuators
  * LCD displays
  * thermo-ceramics. **(4)**

6 Identify the factors a designer might have to take into account when selecting the materials for a door frame. **(5)**

Total for this question paper: 25 marks

Design and technology in society (R402)

This option will be assessed in section B during the $1\frac{1}{2}$ hour, Unit 4 examination. If you have chosen this option, you should spend half of your time (45 minutes) answering all the questions in this section. It is important to use appropriate specialist and technical language in the exam, along with accurate spelling, punctuation and grammar. Where appropriate, you should also use clear, annotated sketches to explain your answer. *You do not have to study this chapter if you are taking the CAD/CAM option or the Mechanisms, Energy and Electronics option.*

Economics and production

You need to

☐ **understand the economic factors of one-off, batch, high volume and continuous production**
☐ **know the sources, availability and costs of materials**
☐ **know the advantages of economies of scale of production**
☐ **understand the relationship between design, planning and production costs**
☐ **understand the material and manufacturing potential for a given design solution.**

KEY TERMS
Check you
understand these terms

variable costs, fixed costs, productivity, scale of production, ore, crude oil, economies of scale, internal/external failure costs

---

 Further information can be found in *Advanced Design and Technology for Edexcel, Product Design: Resistant Materials Technology*, **Unit 4B1, section 1.**

KEY POINTS

Economic factors of one-off, batch, high volume and continuous production

The production chain is the sequence of activities required to turn raw materials into finished products.

Profit is the difference between total costs and income derived from sales. In order to remain profitable, calculations of price must allow for variable costs, fixed costs and a realistic profit. Profits are maximised by reducing costs. Costs are divided into:

- **variable costs**: (direct costs) the costs of production, which vary depending upon the number of products made including materials, services, wages, energy and packaging
- **fixed costs**: (indirect costs or overhead costs) costs, which have to be paid at regular intervals even if production is stopped, including routine marketing maintenance, rent and rates, depreciation of plant and equipment.

Table 4.5 The production chain

Primary sector	The extraction of natural resources as in agriculture, forestry, mining and quarrying. Less economically developed countries (LEDCs) often rely heavily upon the export of raw materials.
Secondary sector	The processing of primary raw materials and the manufacture of products. This sector employs a decreasing proportion of the workforce in the more economically developed countries (MEDCs), such as the USA.
Tertiary sector	The provision of services, which include education, retailing, advertising, marketing, banking and finance. This is a growing sector in MEDCs and the largest employer.

Productivity, labour costs and the scale of production

Productivity is a measurement of the efficiency with which raw materials (production inputs) are turned into products (manufactured outputs), commonly measured as output per worker, or labour costs per unit of production. Productivity is encouraged by:

- setting up an internal market within the company where departments buy and sell their services
- setting budgets for each department.

Table 4.6 Sources of ore

Ore	Sources	Properties
Iron ore	Europe, North America and Australia	High availability, high metal content and relatively low cost
Aluminium ore (bauxite)	Southern hemisphere	Accessible and cheaper to process than steel
Copper ore	Chile, the USA and Canada	Rare and more expensive than iron ore or bauxite

The **scale of production** is important because it affects decisions relating to manufacture, such as the location of factories and the choice of processes. One-off products are more expensive than batch or high volume products due to the high costs of the specialised tools and processes involved, high labour costs and lower levels of productivity.

Sources, availability and costs of materials

Costs of materials

All manufacturers require a reliable supply of raw materials. The sources, availability and costs of materials depend on the type and quantity of materials required. Larger manufacturers, for example, need more materials and can negotiate lower prices. The price of materials increases as a result of limited supply, high demand, complex processing requirements and transportation costs.

Timber, manufactured board, paper, board

The UK imports almost 90 per cent of its timber, which is usually supplied in board form, ready for further processing.

Metals

Metals are extracted from **ores**, which occur naturally in rocks and minerals. Aluminium and iron are the commonest ores, accounting for almost 95 per cent of the total tonnage of all metal production. Smelting ore close to its source often reduces transport and labour costs.

Oil and plastics

Crude oil is untreated oil and is the world's major source of energy, thermoplastics and thermosetting plastics.

The main sources of oil are controlled by the Organisation of Petroleum Exporting Countries (OPEC), a cartel of countries from the Middle East, South America, Africa and Asia, which sets output quotas in order to control crude oil prices. Other oil producing areas include the USA, the Russian Federation and the North Sea. High prices in the 1970s allowed marginal oil fields, such as the North Sea, to be developed.

Plastics are available in sheet and rod or as granules for injection moulding. Plastic resins are supplied in liquid form. Because plastics are widely available, cost effective and require less processing, they have replaced many other materials. For example, PET has replaced glass for bottles.

Advantages of economies of scale of production

Economies of scale are the savings in costs brought about by producing products in larger numbers. One-off products do not benefit from such economies of scale. Products manufactured in high volume and continuous production enjoy lower costs and higher savings. These are brought about by:

- specialisation of labour leading to increased productivity
- the spread of fixed costs
- bulk buying of raw materials at lower unit costs
- lower borrowing costs to fund capital investment
- the concentration of industry, which allows a specialist labour force to develop
- the development of local supply and support networks.

Mass production and the development of new products

The increased pressure to remain competitive forces companies to continually develop new products. The costs of developing new products is high due to the level of investment required. There is a constant need to reduce the time-to-market of new products. The most successful companies produce the right product at the right time, in the right quantity and at the right cost.

The relationship between design, planning and production costs

Total costs and the product's selling price must be set at an appropriate level in order to achieve profit. All costs in a manufacturing company are set in the design phase. Designing for manufacture (DFM) is directly related to designing for cost. The main aims of DFM are minimisation of component and assembly costs, minimisation of product design cycles, and to produce higher quality products.

The cost of quality

Building quality into products incurs costs. However, these costs are usually more than outweighed by the costs of producing poor quality products.

Costing a product

Checks against a competitor's products are often used to establish the potential price range of a new product because setting the price too high

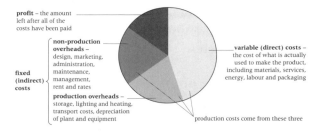

profit – the amount left after all of the costs have been paid

non-production overheads – design, marketing, administration, maintenance, management, rent and rates

fixed (indirect) costs

production overheads – storage, lighting and heating, transport costs, depreciation of plant and equipment

variable (direct) costs – the cost of what is actually used to make the product, including materials, services, energy, labour and packaging

production costs come from these three

Fig. 4.3 The components of a product selling price (SP)

Table 4.7 The cost of quality

The costs of getting it wrong	The costs of checking it is right	The costs of making it right first time
Internal failure costs: Scrap products that cannot be repaired, used or sold; reworking or correcting faults; re-inspecting repaired or reworked products; products that do not meet specifications but are sold as 'seconds'; any activities caused by errors, poor organisation or the wrong materials *External failure costs:* Repair and servicing; replacing products under guarantee; servicing customer complaints; the investigation of rejected products; product liability legislation and change of contract; the impact on the company reputation and image – relating to future potential sales	These costs are related to checking materials, processes, products and services against specifications; that the quality system is working well; the accuracy of equipment	Prevention costs are related to the design, implementation and maintenance of a quality system, including setting quality requirements and developing specifications for materials, processes, finished products and services; quality planning and checking against agreed specifications; the creation of, and conformance to, a quality assurance (QA) system; the design, development or purchase of equipment to aid quality checking; developing training programmes for employees; the management of quality

will constrict sales and setting it too low will reduce profitability. Products have a value, a price and a cost.

- For manufacturers, the product value is always lower than the selling price.
- For consumers, the product value is higher than the selling price.

Profit

Profit is the amount left of the SP after all costs have been paid. Gross profit is calculated by deducting variable plus fixed costs from the sales revenue. Net profit is gross profit minus tax.

Net profit is used to pay dividends to shareholders, bonuses to employees, reinvestment in new machinery, research and development (R&D) of new products and repayment of debts.

The break-even point

A 'break-even analysis' will show the minimum number of units that need to be sold before all costs are covered and profitability is reached.

$$\text{Break-even point} = \frac{\text{Fixed costs}}{\text{Selling price} - \text{variable costs}}$$

Other factors influencing demand, supply and price include:

- what customers perceive as value for money
- what the competition is offering
- how essential the product is to consumers
- political influences
- changes in legislation
- economic conditions
- changes in fashion and trends.

The material and manufacturing potential for a given design solution

Once a design solution has been developed, decisions have to be made relating to pricing, materials, manufacturing processes, quality and quality control, scale of production and size of production runs. The choice of manufacturing process depends upon the quantity of products required and the materials that will be used. The designer will have to adapt any designs to suit the chosen manufacturing process. It is important to plan quality control carefully before manufacturing commences. The production process is often represented as a flow chart showing where and how quality control procedures will be carried out. Sometimes special equipment will have to be made.

EXAMINATION QUESTION

Example question and answer.

Q1 *A manufacturer who does not carry out effective quality control procedures will risk producing faulty products. Describe, in general terms, **three** 'internal failure costs' and **three** 'external failure costs' associated with ineffective quality control and the production of faulty products.* **(6 marks)**

Acceptable answer

Internal failure costs:
Some products may have to be **scrapped** completely, which will cause significant losses. It may be possible to **rework or correct** some of the faulty products but this will take time and money. It may be possible to sell some of the faulty products as **seconds** but the price will have to be reduced, resulting in reduced profits or even losses.

External failure costs:
The manufacturer may have to bear the costs of **repair and servicing** of faulty products. If faulty goods are produced regularly, a reputable manufacturer will have to bear the cost of **servicing customer complaints**. More significantly, the manufacturer may lose the confidence of some consumers, which will affect the **company reputation and image**, threatening future potential sales.

Consumer interests

You need to

☐ **know the systems and organisations that provide guidance, discrimination and approval**

☐ **understand the relationship between standards, testing procedures, quality assurance, manufacturers and consumers**

☐ **know about quality management systems**

☐ **understand the relevant legislation on the rights of the consumer when purchasing goods.**

KEY TERMS

Check you understand these terms

consumer 'watchdog' organisations, *Which?* magazine, ISO 9000, quality management system (QMS), statutory rights

Further information can be found in *Advanced Design and Technology for Edexcel, Product Design: Resistant Materials Technology*, Unit 4B1, section 2.

KEY POINTS

Systems and organisations that provide guidance, discrimination and approval

There are many systems and organisations that provide guidance, discrimination and approval for consumers including:

- the Institute of Trading Standards Administration
- British, European and international standards organisations
- **consumer 'watchdog' organisations**.

Consumer 'watchdog' organisations

General publications, consumer 'watchdog' organisations, specialist magazines and television programmes provide independent consumer advice and information on new products.

- The Consumer's Association publishes the **magazine** *Which?* and a website, which provide reports about product testing and 'best buys', consumer news and links to electronic newspapers.
- The Citizens Advice Bureau (CAB) gives consumer advice and support.
- Product testing and reporting is also carried out by trade and professional organisations and journals.

The relationship between standards, testing procedures, quality assurance, manufacturers and consumers

ISO 9000 is an internationally agreed set of standards for the development and operation of a **quality management system (QMS)**. ISO 9001 and 9002 are the mandatory parts of the ISO 9000 series. They specify the clauses manufacturers have to comply with in order to achieve registration with the standard.

Quality management systems

Quality management systems (QMS) provide an overall approach to ensure that high quality standards are maintained throughout an organisation. A QMS involves a structured approach to ensure that customers end up with a product or service that meets agreed standards. You should be familiar with the quality management system summarised below.

1 Explore the intended use of the product, identify and evaluate existing products and consider the needs of the client.
2 Produce a design brief and specification.
3 Use research, questionnaires and product analysis.
4 Produce a range of appropriate solutions.
5 Refer back to the specification.
6 Refer to existing products. Use models to test aspects of the design.
7 Check with the client. Use models to check that the product meets the design brief and specification.
8 Plan manufacture and understand the need for safe working practices.
9 Manufacture the product to the specification.
10 Critically evaluate the product in relation to the specification and the client. Undertake detailed product testing and reach conclusions. Produce proposals for further development, modifications or improvements.

Applying standards

Risk assessment

As part of the Health and Safety at Work Act 1974, it is a statutory requirement for employers and other organisations to carry out risk assessments in order to eliminate or reduce the chances of accidents happening. There are six steps.

1 Identify the activity or process.
2 Identify potential hazards.
3 Estimate the level and nature of the risk (including people at risk).
4 Establish control measures to reduce risk and decide if the risk has been reduced as far as possible.
5 Eliminate the activity or process or prepare a risk assessment plan.
6 Record assessment and review risk assessment plan.

Ergonomics

Ergonomics makes use of anthropometric data, which exists in the form of charts, and provides measurements for the 90 per cent of the population that fall between the fifth and the ninety-fifth percentiles. Designers make use of this published data to produce better designed products.

Relevant legislation on the rights of the consumer when purchasing goods

Consumers are protected by a body of law called **statutory rights**. You are entitled to have 'reasonable expectations' that manufacturers and advertisers will not mislead you and that their products will not harm you when used in accordance with instructions. You should also expect these products to function properly and to be of sufficient quality to last for a reasonable length of time. Your statutory rights are established by legislation, such as:

- the Sale of Goods Act 1979
- the Supply of Goods and Services Act 1982
- the Sale and Supply of Goods to Consumers Regulations 2002.

The latest 2002 regulations became law in 2003. Consumers now enjoy protection across the EU. The regulations state that:

- every consumer has a right to a repair or replacement if goods are faulty
- for the first six months, it is the responsibility of the retailer to prove that the goods were **not** faulty
- after six months, the consumer has to prove that the goods were faulty (six year limit).

Help in solving problems

Local authority Trading Standards officers enforce and advise on a wide range of legislation relating to consumer protection and deal with problems and complaints.

Table 4.8 Consumers' statutory rights

General statutory rights	These statutory rights protect consumers' 'reasonable expectations' when buying products, irrespective of the supplier. In order to meet the requirements of the Sale of Goods Act, products must satisfy three conditions: • they must be 'of satisfactory quality' and free from defects • they must be 'fit for purpose' • they must be 'as described'. It is the retailer who is responsible, and the customer can expect a refund if the product fails to meet the above standards, as long as it is returned within a reasonable period.
Limits to statutory rights	There are no legal grounds for complaint if consumers: • were told about the fault • did not notice an obvious fault • damaged the product themselves • bought the item by mistake • changed their mind about the product. In order to build customer loyalty, many retailers often exchange products even if they are not faulty and are not under a legal obligation to do so.
Second-hand goods	You have the same rights when buying second-hand goods. You can claim your money back or the cost of repairs if goods sold to you are faulty, provided the faults: • are not due to reasonable wear and tear • were not pointed out to you at the time of sale • were not obvious when you agreed to buy the goods.
Sale goods	You have the same rights when buying sale goods as when buying new ones. Notices that say 'no refunds on sale goods' are illegal.
If things go wrong	If there is something wrong with a product, a consumer should normally get a refund provided the retailer is informed of the problem within a 'reasonable time'. • A consumer should contact the retailer as soon as the fault is discovered and keep a record. • Sellers are legally responsible for the products they sell, not the manufacturers. • Losing the receipt does not mean a loss of statutory rights. • A consumer may be able to claim compensation if a faulty product causes damage. After a 'reasonable time', you may be able to claim compensation, which could take account of the loss in value of the product or a repair or replacement.

EXAMINATION QUESTIONS

Example questions and answers.

Q1 a) *Your friend has statutory rights protected by consumer legislation (laws), such as the Trade Descriptions Act, the Sale of Goods Act and the Consumer Protection Act. Outline* **two** *of these statutory rights enjoyed by your friend and all consumers.* **(2 marks)**

Acceptable answer

Products **must be fit for purpose** so that they can carry out the tasks for which they are designed. It is also illegal to make false claims for any products: they **must be as described**.

b) *Your friend decides to buy a used car from a second-hand car dealer. When the car breaks down, your friend goes back to the second-hand car dealer to ask for a refund. Under what conditions would your friend be legally entitled to a full refund or the cost of repairs?* **(3 marks)**

Acceptable answer

My friend would be entitled to claim back all the money, or be refunded the cost of repairs, if the car was faulty and provided that the faults were **not due to the reasonable wear and tear**, or if they **were not pointed out at the time of sale**, or if they were **not obvious at the time of sale**.

Advertising and marketing

You need to

- ☐ **understand advertising and the role of the design agency in communicating between manufacturers and consumers**
- ☐ **understand the role of the media in marketing products**
- ☐ **understand market research techniques**
- ☐ **understand the basic principles of marketing and associated concepts.**

KEY TERMS

Check you understand these terms

media, advertising agencies, marketing agencies, TMG, test selling, consumer demand, market share, market pull, brand loyalty, lifestyle marketing, competitive edge, product proliferation, price range, promotional gifts

Further information can be found in *Advanced Design and Technology for Edexcel, Product Design: Resistant Materials Technology*, **Unit 4B1, section 3.**

KEY POINTS

Advertising and the role of the design agency

The **media** involves any means of communication such as TV, press and radio, and is often used by advertisers. Advertising relates to **media** communication designed to inform and influence existing or potential customers. Most companies employ **advertising agencies**, which design and plan their campaigns. In marketing, research and planning are used to organise the development and sale of products and services. Although many larger companies have their own marketing departments, most companies use **marketing agencies**, which provide specialist advice and marketing services.

Companies often make use of specialist design and advertising agencies such as Saatchi and Saatchi to promote their products. A 'hard sell' is a simple and direct message that promotes the unique features and advantages of the product (unique selling proposition – USP). A 'soft sell' advertising campaign promotes the personality or image of the product. Brand advertising focuses on creating a positive product image and creating positive emotional and psychological associations with the product. Jamie Oliver's association with the Sainsburys brand is said to have boosted the supermarket's profits by £153 million a year.

Advertising standards

Non-broadcast advertising is regulated by the Advertising Standards Authority (ASA), which checks to ensure that advertisers:

- are legal, decent, honest and truthful
- show responsibility to the consumer and to society
- follow business principles of 'fair' competition.

The role of the media in marketing products

Marketing through the media includes the press, direct mail, broadcast media, cinema advertising, outdoor advertising (billboards and sports grounds), and electronic marketing (direct email and the Internet).

The target froup index (TGI)

The target group index (TGI) is a marketing research organisation that regularly surveys a representative sample of consumers. Subscribing to this survey allows a marketing organisation to match its target market with the media it uses most.

Market research techniques

Market research is expensive, but relying on uninformed decisions is risky and can prove more costly if a product launch is badly planned. Market research is used to identify:

- the nature, size and preferences of current and potential target market groups (**TMG**s) and subgroups
- the buying behaviour of the target market group
- the competition; its strengths and weaknesses (known as competition analysis)
- the required characteristics of new products
- the effect price changes might have on demand
- changes in trends, fashions and consumer tastes (known as trend analysis), such as design, colour, demographics, employment, interest rates and inflation.

Market research comes from two types of sources.

- *Primary sources* provide original research, for example, internal company data, questionnaires and surveys.

Table 4.9 Strengths and weaknesses of the major types of media

Media	Strengths	Weaknesses
Television (around 33% of UK advertising expenditure)	• High audiences, but spread over channels • Excellent for showing product in use	• Short time-span of commercials is limiting • High wastage – viewers not in target market
Newspapers and magazines (around 60% of UK advertising expenditure)	• Can target the market with detailedinformation • Can get direct response (reply coupon)	• Can have a low impact on consumers • Timing may not match marketing campaign
Radio (around 2-3% of UK advertising expenditure)	• Accurate geographical targeting • Low cost and speedy	• Low numbers compared to other media • Listen to it in the background to other tasks
Posters (around 4% of UK advertising expenditure)	• More than 100,000 billboards available • Relatively cheap	• Seen as low impact/ complicated to buy • Subject to damage and defacement

• *Secondary sources* provide published information, for example, trade publications, commercial reports, government statistics, computer databases, the media and the Internet.

Quantitative research collects measurable data, such as sales figures or consumer characteristics. The views of the whole target market group are based on the responses from a sample group. Qualitative research explores consumer behaviour by interviewing individuals about their thoughts, opinions and feelings. It can be used to plan further quantitative research.

Surveys

Typically, surveys are used to collect quantitative data about behaviour, attitudes and opinions of a sample in a target market group. The process can be summarised as follows.

1 Initial exploratory research.
2 Set survey objectives and data requirements.
3 Plan administration of survey.
4 Design questionnaire.
5 Collect data.
6 Process data.
7 Interpret data and write report.

Questionnaires

Questionnaires should be carefully designed to provide useful and accurate information. The questions should be relevant, clear, inoffensive, brief, precise, impartial. Questions can be of two types:

• open (ended) – which allow unlimited responses and are difficult to analyse
• closed questions – which provide a choice of answers and are easier to analyse.

Product analysis

As you have learned from Unit 1 and your coursework, detailed product analysis can help designers to develop detailed specifications and successful ideas.

Test marketing (test selling)

A finished product is placed on sale and data collected to find out customer/retailer reactions.

This data is used to generate forecasts and plan production and marketing strategies. This is known as **test selling**.

Case study: the C5 car

The C5 was launched by Sir Clive Sinclair in 1985 as a new concept in transport. There was very little market research and no test marketing of the C5. When tested by the Consumer's Association, serious criticisms were reported.

• On the road, the C5 was the same height as the bumpers of other cars, which made visibility difficult, increased the chance of accidents, made it vulnerable to exhaust fumes, spray and dazzle from headlights.
• There was no reverse gear so drivers had to get out of the C5 to move it backwards.
• The top speed of the car was 15mph, which caused problems even where the speed limit was 30mph.

Production was stopped after only 14,000 had been produced (compared with an estimated demand of 100,000 per year). Many people put the failure of the C5 down to poor market research.

Examiner's Tip

You will find it helpful to be able to refer to your own case studies that look at the marketing of new and existing products or brands.

The market research process

• Planning: identify a clear reason and purpose (usually a design problem, need or opportunity), decide what data needs to be collected and how to collect it.
• Implementation: carry out the data collection as planned, from primary and secondary sources.
• Interpretation: a report is written to analyse and interpret the collected data. Recommendations are made for future planning.

The basic principles of marketing and associated concepts

Marketing involves anticipating and satisfying consumer needs. Prime marketing objectives include generating profit, developing sales, influencing customers' buying decisions, increasing **market share**, diversifying into new markets, and promotion of company image.

The sales of competing products can be expressed as a proportion of total sales. This is what is meant by market share. Sometimes companies set an objective to increase market share rather than to increase profits. This is especially common when new products are launched and companies are trying to encourage us to switch brands to the new product.

Marketing plan

The basic structure of a product marketing plan includes:

- background and situation analysis, including SWOT (product strengths, weaknesses, opportunities and threats from competition) and PEST (political, economic, social and technological issues)
- information on markets, customers and competitors
- a plan for action and advertising strategies
- planning marketing costs
- time planning; the best time to market the product to an achievable timetable
- a plan for monitoring the marketing.

Target market groups (TMGs)

A market consists of all the customers of all the companies and organisations supplying a specific product. Often it is impossible to supply all potential customers because, for example, they are too scattered or because of strong competition in some areas. As a result, target market groups are identified. These are market segments that have been identified by market research as the most likely potential customers. Companies will find out as much as they can about their TMGs so they can plan their marketing and advertising campaigns to appeal to these customers. The process of identifying TMGs and developing products for them is called target marketing.

Consumer demand and market pull

Consumer demand equates to the number of products sold or projected sales. This is also known as **market pull**. Customers in any market will demand or 'pull' products and services to satisfy their needs. Effective marketing can be used to consolidate **brand loyalty** where consumers make buying decisions based on name and reputation rather than on other factors such as price. Companies use marketing to stimulate demand in order to expand their market share.

Lifestyle marketing

People with similar demographic characteristics often lead similar lifestyles, and demonstrate similar tastes and buying patterns. Profiles are developed that describe the general characteristics of these population groups and **lifestyle marketing** is used to target these potential market groups by matching their needs with products. New products can then be developed to match their needs.

Brand loyalty

A brand is a marketing identity of a generic product that sets it apart from its competitors. Brands are developed over time and marketing strategies aim to associate the brand with attractive images, personalities and emotions. Successful marketing strategies generate strong brand loyalties and consumers are often prepared to pay a significant premium for these products.

Advantages of branding for the customer

Branding provides an expected and reliable level of quality. Strongly branded products can be used as benchmarks and can save time for the consumer when deciding which product to buy. Examples of successful brands include Coca-Cola, Microsoft, IBM, Benetton and Diesel jeans. All these companies have managed to establish a distinct brand image. The consumer perceives these branded products, rightly or wrongly, as better than generic alternatives (e.g. supermarket own brands).

Competitive edge and product proliferation

Price is not the only factor that influences our buying decisions. Manufacturers are always seeking to achieve a **competitive edge** in order to increase sales. This involves incorporating unique qualities or features within products and can be achieved in many ways including:

- price reductions
- higher quality products
- unique features
- enhanced company/brand image through successful advertising.

Successful products, such as Coca-Cola, have become very strong brands throughout the world. **Product proliferation** is achieved when a product becomes nationally/internationally recognised and is consumed across large sections of society.

Price range, pricing strategy and market share

How much consumers are prepared to pay for a product depends on how much they value it. The justification for a higher price may depend on the following:

- a product's extra features, characteristics or innovative design

- the perceived quality of the product
- rarity or shortage of supply
- strong brand image through advertising and promotion
- extra services, such as credit facilities and home delivery.

Sometimes products are sold at a loss to establish or increase market share, to encourage product proliferation or to encourage sales of related products. Manufacturers and retailers establish the **price range** of products. It is expressed as an upper and lower limit and will reflect the unique qualities and features of these products, as well as local market conditions.

Promotional gifts

In some cases, products are given away free. **Promotional gifts** may take the form of samples to encourage people to try a product or they may be products that are designed to strengthen brand awareness.

Distribution

Distribution makes a product available to the maximum number of target customers at the lowest cost. Without an effective distribution strategy, products will not reach the consumer.

EXAMINATION QUESTIONS

Example questions and answers.

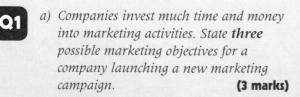 a) *Companies invest much time and money into marketing activities. State **three** possible marketing objectives for a company launching a new marketing campaign.* **(3 marks)**

Acceptable answer
1 Generating **profit**
2 Increasing **market share**
3 **Diversification** into new markets.

b) *Explain, using examples to illustrate your answer, the difference between **one** of the following pairs:*

1 *'Primary sources' and 'secondary sources' of market research data*
2 *'Quantitative research' and 'qualitative research' used in market research.* **(4 marks)**

Acceptable answer (one of the following)
Primary sources of data or information, such as **internal company data, questionnaires and surveys**, produce **original research**. Secondary sources are **existing publications**, such as **trade publications, commercial reports and government statistics**. This form of research is clearly much **less expensive** but **may not provide suitable data**.

Quantitative research collects **measurable data**, such as **sales figures or consumer characteristics**. Qualitative research **explores consumer behaviour by interviewing individuals about their thoughts, opinions and feelings.**

Conservation and resources

You need to

☐ **understand the environmental implications of the industrial age**

☐ **understand the management of waste, the disposal of products and pollution control.**

KEY TERMS

Check you understand these terms

conservation, renewable resources, non-renewable resources, recycling, sustainable development, environmentally friendly, sustainable technology, biotechnology

Further information can be found in *Advanced Design and Technology for Edexcel, Product Design: Resistant Materials Technology*, Unit 4B1, section 4.

KEY POINTS

Environmental implications of the industrial age

Influencing the future

As economies develop, more of the world's resources are being lost and more environments are damaged. Economic activity and levels of consumption are often used as a measure of progress, and twentieth century industry and society has become increasingly wasteful (the 'purchase-attraction' society). In the long term, product designers need to ask how progress can be made in the future by producing fewer products and by making essential products

more environmentally safe. The designer needs to consider issues such as:

- concentrating on the long-term use and usefulness of products
- developing products that remain the property of the manufacturer
- paying for the use of the product and its maintenance
- returning products to the manufacturer to be serviced, repaired, recycled and reused.

Conservation and resource management

Conservation is concerned with the protection of the natural and the man-made world for future use. This includes the sensible management of resources and a reduction in the rate of consumption of finite resources, such as coal, oil, natural gas, ores and minerals. Resources can be divided into:

- **renewable resources** – once consumed they can be replaced, for example, timber
- **non-renewable resources** – once consumed they cannot be replaced, for example, oil.

Efficient management of resources includes:

- using less wasteful mining and quarrying methods
- making more efficient use of energy in manufacturing
- reducing waste materials produced in manufacture
- **recycling** waste materials produced in manufacture
- designing products with components that can be reclaimed and reused or recycled.

In parts of the world, including South East Asia, some tropical hardwoods, such as jelutong, have been harvested faster than the trees can be replaced. In contrast, the Scandinavian timber industry, which provides the majority of paper used in the UK, is arguably one example of **sustainable development.** For each tree that is cut down within these 'managed forests', another is planted; a policy that has led to a net increase in the tree population. An additional benefit is the fact that young trees absorb more carbon dioxide and release more oxygen, helping to combat the greenhouse effect.

Renewable sources of energy, energy conservation and the use of efficient manufacturing processes

There has been an increasing reliance on non-renewable sources of raw materials and energy, such as coal and oil. Consumer pressure, cost reduction and national/international legislation have forced product designers to consider:

- reducing the amount of materials used in a product
- reusing products and waste materials within a manufacturing process
- recycling waste in a different manufacturing process
- using efficient manufacturing processes that save energy and prevent waste
- designing for easy product maintenance
- designing the product so that the whole or parts of it can be reused or recycled.

Turning to renewable sources of energy is one means of conserving non-renewable resources.

The use of efficient manufacturing processes

Product manufacturers can contribute to sustainable development and cost reduction by using more efficient design and manufacturing processes.

Table 4.10 Renewable sources of energy

Source	Process	Advantages	Disadvantages
Wind	Power of wind turns turbines	Developed commercially Produces low-cost power	High set-up cost Contributes a small proportion of total energy needs Wind farms sometimes seen as unsightly
Tides	Reversible turbine blades harness the tides in both directions	Occurs throughout the day on a regular basis Reliable and non-polluting Potential for large-scale energy production	Very high set-up cost Could restrict the passage of ships Could cause flooding of estuary borders, which might damage wildlife
Water	Running water turns turbines and generates hydro-electric power	Clean and 80-90% efficient	High set-up cost Suitable sites are generally remote from markets Contributes to a small proportion of total energy needs of an industrial society
Solar	Hot water and electricity generated via solar cells	Huge amounts of energy available. Could generate 50% of hot water for a typical house Relatively inexpensive to set up	High cost of solar cells Biggest demand in winter when heat from Sun is at its lowest
Geothermal	Deep holes in Earth's crust produce steam to generate electricity	Provides domestic power and hot water	Only really cost-effective where Earth's crust is thin, e.g. New Zealand, Iceland

Case study: Yellow Pages

Yellow Pages is used by almost 27 million households. A recent design overhaul led to savings totalling £1.5 million per edition and a large reduction in the quantity of materials used. The changes in design were popular with customers who found the book easier to use, which led to an increase in the number of advertisers buying space. The edition included:

- an emphasis on the Yellow Pages brand
- a new cover design, a smaller but clearer font and a more compact layout
- an index reduced from twenty-three pages to seven
- an emphasis on the local character of each edition, using customer friendly cartoon-style images.

New technology and environmentally friendly manufacturing processes

The government agency, Envirowise, aims to help manufacturing companies improve their environmental performance and increase their competitiveness. **Environmentally friendly** processes reduce damage to the environment and the depletion of non-renewable resources. The main themes of the programme are waste minimisation, which reduces costs, and the use of cleaner technology, reducing the consumption of raw materials, water and energy.

Reducing waste in the paper and board industry

A large number of additives are used to improve machine performance and paper quality. These non-fibrous materials include cleaning chemicals for papers and machines, fillers to improve paper opacity, sizing agents, dyes, optical brightening agents, and paper coating chemicals. The overuse of these additives is a result of:

- a lack of operating manuals
- specifying the wrong type and dose of chemicals
- using incorrect sizes of dosing pipes
- using incorrect pump speeds
- incorrect connections between chemical storage systems and papermaking machinery

- poor labelling on storage tanks, pumps and pipe-work.

Reducing these additives by one per cent could save the industry in the region of £4 million per year. The improved management of non-fibrous materials can:

- reduce raw materials costs
- reduce the generation of waste
- reduce waste disposal costs
- increase the amount of saleable paper produced
- reduce machine down time
- reduce production losses by up to five per cent.

Cleaner technology using high volume, low pressure (HVLP) spray guns

In order to judge the true environmental impact of any product, it is important to study the whole life cycle of that product.

Case study: HVLP spray guns

Compared with conventional spray guns, HVLP guns atomise paint using a higher volume of air at a lower pressure. They reduce solvent use and meet the requirements of the 1990 Environmental Protection Act. The benefits of the use of modern HVLP spray guns include:

- an initial reduction in paint use of up to twenty-one per cent with obvious cost savings
- reduced use of solvents, compressed air and energy, further reducing costs
- a short payback period on the purchase price of the spray guns
- reduced environmental impact
- a better finish as over-spraying is reduced.

The importance of using sustainable technology

Sustainable technology and sustainable development are philosophies that emphasise an environmentally friendly approach to technological and economic development. Economic growth is welcomed, but people are encouraged to see the environment as an asset, which should be shared and protected by all nations and future generations. Sustainable development has been described as:

'development that meets the needs of the present, without compromising the ability of future generations to meet their own needs.'

The key objectives of sustainable development include:

- giving priority to the essential needs of the world's poor
- meeting essential needs for jobs, energy, water and sanitation
- ensuring a sustainable level of population
- conserving and enhancing the resource base
- bringing together the environment and economics in decision making.

'Bio-Wise' is a government initiative that supports and advises companies and organisations on developing sustainable practices that make use of **biotechnology**.

Management of waste, the disposal of products and pollution control

There are three key approaches to reducing waste (the three Rs):

- reduce the amount of materials used in manufacture
- reuse materials in the same manufacturing process where possible
- recycle materials in a different manufacturing process if possible.

Reducing materials use

Small changes to manufacturing processes can lead to a significant reduction in waste and costs.

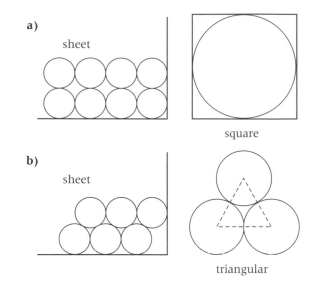

Fig. 4.4 Maximising the use of aluminium sheet in the production of aluminium can tops: arrangement A produces 21.4 per cent waste, while arrangement B produces only 9.3 per cent waste

Pollution is the environmentally harmful by-products of industrial activity. In the UK, the 1990 Environmental Protection Act (EPA) introduced tight controls on the discharge of waste into water, land and air. Prosecution can lead to large fines and expensive clean-up costs.

Impact of biotechnology on manufacture

Biotechnology is the use of biological processes involving live organisms in industrial processes. Traditional applications include the use of yeast to make bread and enzymes added to improve the properties of laundry detergents. Consumer pressure, cost advantages and legislation have encouraged companies to develop and apply new, environmentally friendly biotechnologies.

Table 4.11 The disposal of products and pollution control

Destination	Issues
Landfill 90%	Some materials, such as many plastics, will not decompose. Materials that do decompose release harmful chemicals and explosive gases, which need to be managed. Landfill sites are unpleasant and use large areas of land.
Incineration 5%	Regarded as a more environmentally friendly method. However, burning some materials can release harmful gases, which need to be filtered out. The process is more expensive than landfill, but some incinerators can recoup costs by sorting and recycling materials and converting the heat generated into electricity.
Recycling 5%	Some countries recycle a large proportion of household waste. However, recycling can be a costly process and while it conserves resources, the process also consumes power and resources. Hence the need to 'design for recycling'.

These include:

- reed beds and aerobic and anaerobic treatment of industrial effluent
- composting to treat domestic, industrial and agricultural organic waste
- 'bioremediation' techniques to clean up contaminated land
- the treatment of waste cutting fluids.

Turning waste wood dust into garden compost

It is now possible to turn hazardous wood dust, a by-product of timber and manufacturing industries, into safe garden compost. This has been made possible by the discovery of microbes that feed on the lacquers, sealers and solvents used in these industries. Computers manage a controlled environment, which optimises the action of these micro-organisms. This process reduces pressure on landfill sites and allows industry to generate a return on a previously useless by-product.

Biochips

The number of circuits that can be placed on a conventional integrated circuit (IC) is limited by:

- the width of the circuit tracks
- the occurrence of short circuits
- the heat produced by large numbers of components.

Biochips have been developed to replace silicon chips with semi-conducting molecules in a protein framework. The proteins are grown to take up the complex 3D structures, which can eventually lead to further miniaturisation in electronic products.

The advantages and disadvantages of recycling materials

The suitability of materials for recycling depends upon:

- the value of the materials when recycled
- the costs of processing.

Metal, glass and paper are the most cost-effective materials to recycle and are treated differently according to their value.

- Non-ferrous metals are sorted into grades due to their relatively high commercial value.
- Steel and cast iron are graded by size due to their lower commercial value.
- Items made of glass, rubber and plastics are more difficult to sort and have a much lower commercial value than metals.
- Plastics, paper and glass recycling are growing industries.
- Waste materials from the manufacturing process are the most valuable because their material content is known and they are easily available.
- Waste from old products is more difficult to process due to the varied nature of the chemical or physical make-up.

The advantages for the environment of recycling include:

- conservation of non-renewable resources and reduced dependency on raw materials
- reduced energy consumption
- less pollution, including greenhouse gas emissions.

Design for recycling

Due to existing and proposed European regulations, manufacturers are being encouraged to take more responsibility for the recycling of products. As an example, some products are designed to disassemble themselves under controlled conditions. UK engineers have developed a mobile phone that falls apart when heated. The phone is made from shape memory polymers, plastics that revert to their original form when heated. Different components of the phone will change shape at different temperatures, allowing them to fall off at different times when they pass along a conveyor belt. The reusable parts can then be recycled.

Recycling packaging

Facts and figures

- 1.7 million tonnes of plastic waste is produced per year in the UK.
- Less than fifteen per cent is recycled.
- Around 85 per cent goes to landfill where it remains for a hundred years or more.

- An EC Directive sets a target of at least 50 per cent of plastic packaging to be recycled.

Sorting plastics

Recycling is a complex and time-consuming process, but it has become more efficient with the increase of packaging designed for recycling with:

- easily removable labelling
- material identification symbols moulded into products.

Recycling is a growing concern for the packaging industry and local government. However, many local councils currently do not have adequate recycling facilities beyond those used for paper, aluminium and glass. As a result, tonnes of plastic packaging still ends up in landfill. Plastics can be identified manually by using their identification symbols, but it is more cost effective to use automated methods, for example:

- as HDPE and LDPE floats, they are retrieved by placing the waste in a water tank
- PET and polyvinyl chloride (PVC) sink and are identified chemically using sensors.

EXAMINATION QUESTIONS

Example questions and answers.

 Q1

a) *Discuss the aims and objectives of sustainable development.* **(5 marks)**

Acceptable answer

Sustainable development **encourages environmentally friendly economic development**. The Earth, its environment and its resources should be treated as **an asset** for this generation, which **requires careful management and investment for future generations**. Sustainable development is a global philosophy in which **priority is given to the essential needs of the world's poor**. It is important to **limit the world's population** to a level that can be supported by available resources. When governments or international organisations make decisions about development, they are encouraged to **incorporate environmental costs into economic calculations**.

b) *Explain the term 'biotechnology' and give* two *examples of its use.* **(3 marks)**

Acceptable answer

Biotechnology is the **use of biological processes involving live organisms within industrial processes**. Biotechnology has been used successfully in the **composting processes used to treat domestic, industrial and agricultural organic waste**. Biological organisms have also been used **to treat waste cutting fluids from industrial machinery**.

PRACTICE EXAMINATION STYLE QUESTION

1 a) Crude oil is a non-renewable resource. Explain the term 'non-renewable'. **(1)**

   b) Raw materials come from many different sources. Name the main producers of crude oil and outline its importance as a source of other raw materials. **(3)**

   c) Explain how industry can help to conserve fossil fuels and other non-renewable resources:
  * during the design stage
  * during manufacturing. **(5)**

   d) Discuss the issues surrounding the disposal of waste products when they reach the end of their life. **(6)**

2 a) Marketing relies on the media. Explain the terms:
  * marketing
  * media. **(2)**

   b) Describe the role of the Advertising Standards Authority (ASA) in the advertising industry. **(3)**

   c) Discuss the importance of branding and brand loyalty with reference to a product or products of your choice. **(5)**

   d) Surveys and questionnaires are frequently used to gather valuable research. Outline the main stages of the process followed when creating and conducting a customer survey. **(5)**

Total for this question paper: 30 marks

This option will be assessed in section B during the $1\frac{1}{2}$ hour, Unit 4 examination. If you have chosen this option, you should spend half of your time (45 minutes) answering all the questions in this section. It is important to use appropriate specialist and technical language in the exam, along with accurate spelling, punctuation and grammar. Where appropriate, you should also use clear, annotated sketches to explain your answer. *You do not have to study this chapter if you are taking the Design and Technology in Society option or the Mechanisms, Energy and Electronics option.*

Computer-aided design, manufacture and testing (CADMAT)

You need to

- ☐ understand computer-aided design, manufacture and testing (CADMAT)
- ☐ understand computer integrated manufacture (CIM)
- ☐ understand flexible manufacturing systems (FMS)
- ☐ understand CADMAT, FMS and CIM applications within:
  - ☐ creative and technical design
  - ☐ modelling and testing
  - ☐ production planning
  - ☐ the control of equipment, processes, quality and safety
  - ☐ the control of complex manufacturing processes
  - ☐ integrated and concurrent manufacturing.

Further information can be found in *Advanced Design and Technology for Edexcel, Product Design: Resistant Materials Technology*, Unit 4B2, section 1.

KEY POINTS

Computer-aided design, manufacture and testing (CADMAT)

Manufacturers have always experienced pressures to maximise profit and reduce costs. Global manufacturing places additional pressures on manufacturers who have to:

- compete with low-cost imports
- satisfy customer demands for shorter production runs
- satisfy customer demands for improved delivery times.

Computer-aided design, manufacture and testing (**CADMAT**) is an extension of CAD/CAM and fully integrates the use of computers at every level and stage of manufacturing. Computers are used to manage

KEY TERMS
Check you understand these terms

CADMAT, PDM, JIT, CIM, FMS, TQM and TQ, concurrent manufacturing

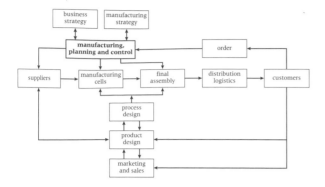

Fig. 4.5 A system flow chart describing integrated manufacturing

data and to help the decision-making process in a variety of ways:

- gathering, storing, retrieval and organisation of data, information and knowledge
- computer model simulations to test new production methods or systems
- mathematical analysis of designs
- communication between project team and clients
- process and equipment control including the scheduling of routine maintenance
- monitoring safety and quality.

Figure 4.5 is a simplified model showing the complex relationships in the design and manufacturing process. Failure in one area can hold up the entire process and it is vitally important that up-to-date, 'real time' data is immediately available so that the system can be managed effectively.

Project data management (PDM)

PDM provides a system to centralise and manage all data generated from the design and manufacturing process. Data is gathered from all areas and stored on a secure database. Different levels of access are established according to need and PDM also allows rapid communications between departments, manufacturers, retailers and customers, who can track their orders online.

The advantages of PDM include:

- fewer bottlenecks
- improved quality control
- more accurate product costing
- more effective production control

- making it easier to work at optimum capacity with minimum stock levels
- increased flexibility
- higher productivity and profits.

Just in time (JIT)

Companies are always seeking ways of reducing costs. Maintaining high stock levels incurs significant costs.

- Capital is tied up in raw materials, components, sub-assemblies and unsold products.
- Too much stock at the point of production can create safety issues and lead to deterioration.
- Storage of raw materials, components, sub-assemblies and unsold products costs money.

This is particularly significant for high volume manufacturers. On the other hand, stock shortages in one section of the production line can halt production altogether. **JIT** is a management philosophy that seeks to minimise these costs. In the automotive industry, some components are manufactured by suppliers literally hours before they are required at the factory. They are delivered to the production line so that each component arrives in the correct order. The result is that each product is manufactured to the specification of an individual customer.

Aims of JIT

- Raw materials, components and sub-assemblies are received from suppliers just before they are needed.
- Quick response to customer orders.
- Goods are supplied to a clearly defined level of quality and quantity.
- Waste (raw materials, time and resources) is minimised.
- Individual manufacturing units throughout the organisation follow the JIT philosophy as well as manufacturers, suppliers and customers.

Features of JIT

- Continuous operation.
- All employees are responsible for quality.

- Manufacturing is synchronised to avoid bottlenecks.
- 'Kaizen' – continuous improvement is encouraged to increase efficiency and reduce waste.
- Work is simplified using foolproof ('poka-yoke') tools, jigs and fixtures.
- The end product is the important focus and anything that does not contribute toward this should be removed.
- Factory layouts should be custom designed to minimise movement of materials and products.
- 'Jidoka' – automation seeks to use self-regulating machines, reducing the need for direct human intervention and allowing decisions to be made centrally, based on electronically gathered data.

JIT and workplace organisation

- Operational set-up times are reduced, increasing flexibility and capacity to produce smaller batches.
- A flexible, multi-skilled workforce is required, which leads to greater productivity and job satisfaction.
- Production rates are flexible and can be levelled or varied, smoothing the flow of products through the factory.
- 'Kanban' control systems are used to control and schedule production rates. (Kanbans originally were cards that identified components, authorised and recorded operations, and followed batches of components around the factory.)

Computer integrated manufacture (CIM)

A **CIM** system uses ICT to integrate all aspects of a company's operations (production, business and manufacturing information) in order to create more efficient production lines. Tasks within a CIM system include :

- the design of components
- planning the most effective production sequences and workflow
- control of machining operations
- performing business functions such as invoicing customers and ordering materials.

Flexible manufacturing systems (FMS) and their wider application in industry

Characteristics of **FMS** include:

- the system responds quickly to changes in demand or supply
- multi-purpose equipment and techniques are used to increase operational flexibility
- system effectiveness is constantly monitored and evaluated
- lead times from design to manufacture are reduced
- modifications to designs are incorporated rapidly
- minimum stock levels
- close relationships between suppliers, manufacturers and retailers (vertical partnerships)
- increased sales and stock turnover
- use of sophisticated computer-based management tools, such as manufacture resource planning (MRP), provide real time processing data to track work through production cycle and to re-plan production in response to changes in demand.

Creative and technical design

Computer-aided engineering (CAE) supports and overlaps CAD/CAM technology. CAE takes advantage of ICT technology to gather, analyse and manage engineering data. For example, computers are used to simulate different conditions to test product performance. This is a convenient and relatively inexpensive technique, which shortens the time needed for product development.

Modelling and testing

Modelling techniques have been covered in earlier units. Look at Unit 3B2 Computer-aided design, The use of CAD to aid the design process; Unit 3B2 Computer-aided manufacture, The applications, advantages and disadvantages of common CNC equipment.

Examples include:

- experimenting with different 3D forms and layout

- testing models using specialist software, for example, manufacturing processes can be simulated to test suitability of the design for production
- virtual interactive models and environments are created using virtual reality modelling language (VRML).

Production planning

FMS and similar systems allow products to be produced in any order using any available machine. Careful and flexible planning is needed to make full use of the resources – 'time is literally money'. Any strategies that save time or increase responsiveness will increase profitability and generate a return on the high level of investment. Scheduling is part of the planning process and is supported by a wide range of software tools. Scheduling seeks to specify the scope and detail of work, the production start date, the production deadline, the machinery and processes required, and labour requirements.

There are two categories of scheduling.

- *Finite capacity* scheduling bases planning on available production capacity.
- *Infinite capacity* scheduling bases planning on customer deadlines assuming that sufficient capacity will always be available. A master production schedule (MPS) is used in infinite capacity scheduling systems to set short-term production targets based on known demand, forecasts and planned stock levels.

The advantages of time-based strategies over cost-based strategies

Pressures created by cost-based strategies can lead to low wages, narrow product focus, relocation to low-wage economies, and centralisation. Time/FMS/rapid response-based strategies are not as vulnerable to such pressures and are more inclined to site themselves close to customers.

Types of computer-aided scheduling functions

- *Resource scheduling* is a finite scheduling function that focuses on the resources needed to turn raw materials into products.

Data from sales orders, stock recording and cost accounting is entered into a database, but the reliability of the system relies upon the accuracy of data input.

- *Electronic scheduling boards* replace card-based boards automating manual processes, such as estimating production time, and provide notification of problems or conflicts.
- *Order-based scheduling* prioritises the distribution of parts and components.
- *Constraint-based schedulers* locate potential bottlenecks and ensure that these are well provided.
- *Discrete event simulation* simulates the production line point by point to resolve bottlenecks and other production problems.

Control of equipment, processes, quality and safety

Control and feedback systems

Equipment can be controlled manually or by electrical, electronic, mechanical, pneumatic, computer or microprocessor systems. Control systems depend on feedback to maintain efficiency, accuracy, reliability, safety and minimal levels of waste. Control systems are found in:

- *materials handling:* for example, ensuring materials arrive in the right place at the right time
- *materials processing:* for example, equipment will automatically respond to changes in conditions; feeds and speeds will be adjusted automatically
- *joining materials:* for example, heat sealing and bonding of plastics for PoS displays using electronic or computer control
- *quality control:* for example, optical monitoring of ink quality adjusts feed rates
- *safety systems:* for example, machines will not operate unless guards are in place or operators are out of the way
- *co-ordination of production:* for example, graphical display panels will show active cells of equipment and sound alerts when problems are identified.

Total quality (TQ)

Total quality systems, such as total quality control (TQC) and total quality management

(**TQM**), link together quality assurance and quality control into a coherent improvement strategy. **TQ** programmes are purpose driven, long-term processes that engage everyone in the design and manufacture process and involve comprehensive change.

ICT forms an important part of TQ processes including interactive multimedia support for employee training and computer-aided statistical tools and methods for checking quality. For example, product barcodes can be used to trace manufacturing defects through a PDM system.

There are three stages in the development of TQ.

1 Raising awareness
2 Empowerment
3 Alignment

TQC and TQM tools

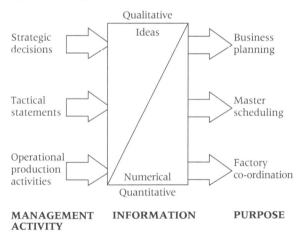

Fig. 4.6 The relationship between thinking tools and planned outcomes

Statistical process control (SPC) and the use of ITC

The study of statistics is concerned with the collection, analysis, interpretation and presentation of numerical data. The complexity of modern manufacturing organisations results in a large amount of data that necessitates the use of modern statistical techniques including pareto charts, flow charts, cause and effect/fishbone/Ishikawa diagrams, bar graphs and histograms, check sheets, and checklists.

Monitoring and inspecting quality

Despite QA procedures designed to eliminate faults before they happen, inspection programmes are still important. The costs associated with undetected faults escalate as the product moves through the production process and on to the customer.

- *Manufacturing cells* – each team or cell is responsible for product quality and has to meet specified quality indicators.
- *Artificial vision* – manual inspection methods are dependent on the effectiveness of the individual. Optical print monitoring QC systems:
  - high speed, real time fault detection
  - automatic feedback and solutions
  - labour relieved to perform more productive tasks
  - lower operational costs
  - quality specifications can be adjusted.

Manufacturing to tolerances

- Tolerance is the 'margin of error' or degree of imperfection allowed in a product or component.
- Tolerance limits are expressed as a +/– figure.
- Tolerances can be checked manually through visual inspections or by using sophisticated probes and sensors including laser technologies.
- The data is gathered as on-process or post-process measurements and analysed.
- Tolerances can be set for any property including size, weight, colour, strength.

Control of complex manufacturing processes

Computer systems are used to:

- optimise factory layout of plant and equipment
- provide effective deployment of labour
- schedule processing operations
- monitor and control workflow
- manage and disseminate production data.

Managing workflow

Production plans and control systems manage the input of materials and components, work in progress, and output to ensure orders meet deadlines to cost.

Monitoring workflow

Laser and barcode (data recognition tags) technology is used to record and monitor the progress of production, sending data to the supervising computer controlling the production line.

Controlling workflow

Project management software co-ordinates production cells within a Master Production Schedule (MPS). Workflow software ensures that individuals responsible for particular tasks are notified and provided with the information necessary to complete the tasks.

Integrated and concurrent manufacturing

Sequential manufacturing

Product development follows a linear path. Each stage of a product life cycle is dependent upon the completion of all preceding stages. This system has several disadvantages.

- In order to trace and correct faults, the product has to be sent back through each preceding stage.
- It is slow to respond to change or fluctuations in demand.
- There are longer lead times.

- There are quality problems due to separation and isolation of each department and costly redesign loops.

Concurrent manufacturing

Concurrent manufacturing (simultaneous manufacturing) is applicable to batch and volume manufacturing. Features and advantages of concurrent manufacturing include:

- operations run alongside each other
- team-based approach
- all departments are represented to ensure quality decision making
- shorter lead times
- manufacturers are forced to consider the whole product life cycle from conception to disposal
- suppliers and retailers involved early in the product development cycle
- good communications are essential, increasingly taking advantage of EDI systems
- the use of ITC becomes more important in design for manufacture (DFM)
- creation of customised 'expert systems' shared on company intranets
- the Internet is used to allow access to information across the world
- enables the use of JIT and quick response manufacturing
- production deadlines (milestones) become the responsibility of individual team members.

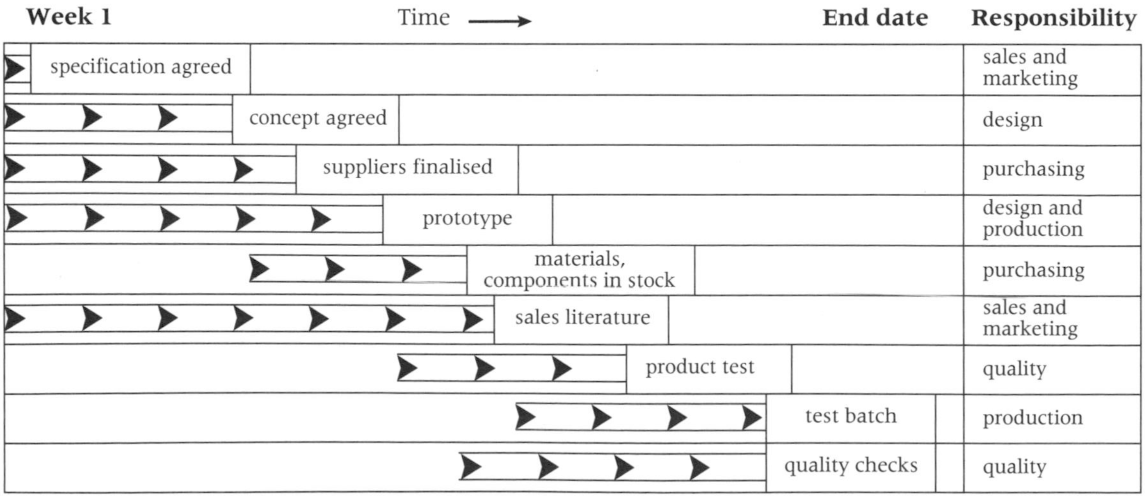

Fig. 4.7 A simplified milestone plan (Gantt chart) for a typical manufactured product

EXAMINATION QUESTIONS

Example questions and answers.

Q1

a) *Explain how computers are used to organise data and help the decision-making process in computer-aided design, manufacturing and testing (CADMAT).* **(4 marks)**

Acceptable answer

CADMAT is an extension of CAD/CAM and **fully integrates the use of computers at every level and stage of manufacturing**. Computers are used to **gather, store, retrieve and organise data, information and knowledge**. Testing can be carried out using **computer model simulations to prototype new production methods or systems**. Designs can be assessed much more thoroughly using **mathematical analysis**. **CADMAT** systems allow **efficient communications** between project teams and clients. Process and equipment control, such as the **scheduling of routine maintenance**, can be organised efficiently to avoid impacting production. Computer systems are used to **monitor safety and quality**.

b) *Just in time (JIT) is a management philosophy that seeks to maximise efficiency and minimise costs. Explain the two most significant costs that are minimised by JIT systems.* **(2 marks)**

Acceptable answer

Capital costs, which are tied up in raw materials, components, sub-assemblies and unsold products.

Storage costs of raw materials, components, sub-assemblies and unsold products costs.

c) *Describe some of the main aims and/or features of JIT and its effect on workplace organisation.* **(7 marks)**

Acceptable answer

Raw materials, components and sub-assemblies are only **delivered to the production line just before they are needed**. Every level of the manufacturing organisation, including **individual manufacturing units, suppliers and customers throughout the organisation, follow the JIT philosophy**. JIT systems are able to provide a **quick response to customer orders** and **goods are supplied to a clearly defined level of quality and quantity**. JIT is a **continuous operation, which synchronises manufacturing to avoid bottlenecks. A flexible, multi-skilled workforce is required**, which leads to **greater productivity and job satisfaction**.

Robotics

You need to

- [ ] **know the industrial application of robotics/control technology and the development of automated processes**
- [ ] **understand complex automated systems using artificial intelligence (AI) and new technology**
- [ ] **understand the use of diagrams to represent simple and complex production systems**
- [ ] **know the advantages and disadvantages of automation.**

KEY TERMS

Check you understand these terms

robotics, artifical intelligence (AI), ASRS, AGVs, expert systems, NLP, neural networks, fuzzy logic, machine vision systems, voice recognition systems, open- and closed-loop control, EDI

Further information can be found in *Advanced Design and Technology for Edexcel, Product Design: Resistant Materials Technology*, Unit 4B2, section 2.

KEY POINTS

The industrial application of robotics/control technology and the development of automated processes

Table 4.12 Robotic and control technologies

Technologies	Definition
Automation	The automatic operation and self-correcting control of machinery or production processes by devices that make decisions and take action without the interference of a human operator.
Robotics	A specific field of automation concerned with the design and construction of self-controlling machines or robots.
Mechatronics	Mechatronic devices integrate mechanical, electronic, optical and computer engineering to provide mechanical devices and control systems that have greater precision and flexibility.
Animatronics	Animatrons are animated sculptures used to entertain or inform as you might see at theme parks.
Artificial intelligence (AI)	Devices that have the capacity to learn from experience and which seek to imitate aspects of human thought processes.

Applications of robotics

The original driving forces behind the development of robotics technology were military and automotive applications. Typical roles and environments have been described as those which are dirty, dull, dangerous, demeaning, hot, heavy or hazardous.

Increasingly, robots have been applied to dangerous and/or repetitive tasks throughout

Fig. 4.8 Canadarm: a robot arm used on the space shuttle

many industrial sectors, such as assembly, painting, palletising, packing, welding, dispensing, cutting, laser processing, and material handling. Other emerging applications are in the advertising, promotional, leisure and entertainment industries, such as animatronics. Robots are slowly finding their way into warehouses for automatic stock control, laboratories, research and exploration sites, energy plants, hospitals, and even hostile environments, such as outer space.

Basics of robot design

For a machine to be classified as a robot, it usually has five parts (Table 4.13).

Work envelope

The work envelope is the range of space in which a robot can work; the maximum limits of movement of the end-effector in all directions. The work envelope is dependent upon the range of movement of the robot: jointed, cylindrical, spherical, or Cartesian (rectilinear). Robots may be fixed, move along overhead rails (gantry robots), or move along the floor (mobile robots).

Basics of robot design

For a machine to be classified as a robot, it usually has five parts.

Table 4.13 The parts of a robot

Controller	The controller is the 'brain' of the robot and allows it to be networked to other systems. Controllers are run by programmes written by engineers. In the future, AI robots will be able to adapt their programming in response to their environment.
Arm	The arm is the part of the robot that positions the end-effector and sensors. Each joint is said to give the robot 'one degree of freedom'. Most working robots today have six degrees of freedom.
Drive (actuators)	The drive is the 'engine' that drives the links (the sections between the joints) into their desired position. Air, water pressure or electricity powers most drives (actuators).
End-effector	The end-effector (end of arm tooling) is the 'hand' connected to the robot's arm. It could be a tool (or interchangeable tools) such as a gripper, vacuum pump, tweezers, scalpel, blowtorch or heat-sealing gun.
Sensor	Feedback is provided by sensors which send data back to the controller which will make adjustments in response to changes in the robot's environment.

The role of robots in batch and volume production

- Robots support continuous production that involve many repetitive tasks.
- Robots encourage flexibility because they can be reprogrammed to perform different tasks.
- They allow many processes to run concurrently, sequentially or in combination.
- A limited number of human operators are required.
- Safe and secure 'fail-safe' operating conditions use sensors to shut down operations in the event of a failure or emergency.

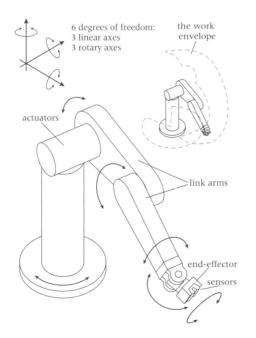

Fig. 4.9a A robotic arm

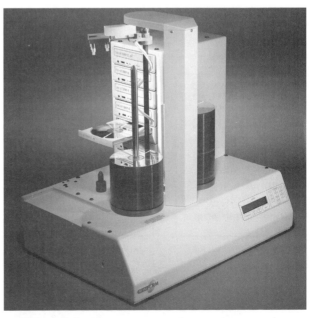

Fig 4.9b A CD batch printing system employing a robotic arm to locate CDs

These robotic technologies are enabling technologies that allow innovative solutions to complex processing operations.

Control systems

Robots use sensors (transducers) to detect or measure properties. Computer or microprocessor control systems record and display this data, then regulate, check, verify or restrain actions of the robot or automated system.

The data generated by these transducers is processed within the robot control system, often using programmable logic controllers (PLCs), in order to perform tasks, or to be recorded, stored and sent to a display device, which can be monitored by the operator.

Monitoring quality and safety

- Data from individual robots, machines, and production cells is gathered locally and centrally. This data is analysed to identify problems before they affect production.
- The development of modern sensing techniques allows effective automated online inspection.

- Quality checks are conducted at each stage of manufacture before components are moved on to the next stage.
- 'Fail-safe' procedures are built into automated machinery to eliminate accidents or poor quality components.

Types of robots include:

- simple robots that can be programmed to perform specific and repetitive tasks, for example, numerically controlled robots, which are programmed and operated much like CNC machines
- remote and sensor controlled robots that use data or commands generated outside their internal programming.

Automatic storage and retrieval systems (ASRS)

- **ASRS** are used in automated warehouses.
- They are commanded externally either to transfer an object from a designated pick-up point to a designated position in storage or vice versa.
- The required motions are internally programmed into the device.

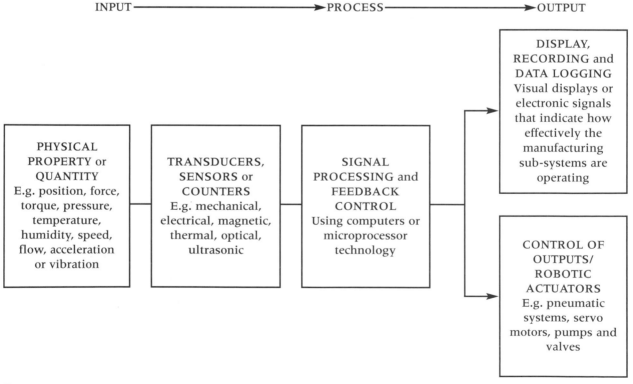

Fig. 4.10 A basic control system

- They are available in various sizes.
- The commands may come from a computer, which controls a larger operation, in which case the robot computer and the computer that commands it are said to form a control 'hierarchy'.

Mobile robots or automated guided vehicles (**AGVs**) are unmanned vehicles that carry 'loads' along a pre-programmed path and are typically used in component or pallet transfer. AGVs use different navigational systems and travel around under a combination of automatic control and remote control. Other uses of AGVs are in mail delivery, surveillance and police tasks, material transportation in factories, internal pipe inspection, military tasks, such as bomb disposal, and underwater tasks, such as inspecting pipelines and ship hulls and recovering torpedoes.

New uses for robots in manufacturing

Robots are increasingly used in the plastics manufacturing industry.

- Finishing plastic components using a six axis robot with a trimming device for drilling, routing, blow moulded bubble removal, flash removal, de-gating or any other type of plastic removal application.
- Insert loading systems in which the robot will load metal threaded inserts, other plastic components and appliqués into the moulds.
- Part removal systems, which remove parts from the moulding machines.
- Infrared plastic welding systems.
- Laser cutting systems.

The key benefits of robotics are:

- continuous operation – most robots can switch effortlessly between different pre-programmed operations without the need to stop production
- reproducibility – pre-programmed operations can be stored indefinitely until required
- consistent quality – the accuracy and reliability of robots and automatic sensors virtually eliminates human error
- safer work environment – the ability of robots to operate in hazardous environments

combined with 'fail-safe' procedures leads to a reduction in risk to human operators
- reduction of labour costs – robotics requires heavy initial investment but automated systems have much lower operational costs.

Complex automated systems using artificial intelligence (AI) and new technology

AI seeks to imitate characteristics associated with human intelligence (such as learning, reasoning, problem-solving and language). In short, AI seeks to create devices that can 'think'. In 1997, an IBM supercomputer called 'Deep Blue' defeated the world chess champion, Garry Kasparov.

AI activities broadly encompass:

- **expert systems**
- computer/machine vision
- natural language processing (**NLP**)
- artificial **neural networks**
- **fuzzy logic**.

Features of a thinking machine

- It must be able to perceive and understand.
- It must possess intelligence and knowledge including the ability to solve complex problems or make generalisations and construct relationships.
- It must be able to consider large amounts of information simultaneously and process them faster in order to make rational, logical or expert judgements.
- It has to pass the Turing test, which states that a computer would deserve to be called intelligent if it could deceive a human into believing that it was human.

Knowledge-based or expert systems

Expert systems are designed by knowledge engineers who study how experts make decisions. They identify the 'rules' that the expert has used and translate them into terms that a computer can understand. This is then stored as a knowledge base, which can be used to solve real-life problems.

Application of AI in design and manufacture

AI is still a developing technology that is not widely used, decisions being made by product development teams. However, elements of AI are available to be used in:

- CAD, process planning and production scheduling
- problem diagnosis and solution in machinery and equipment
- modelling and simulation of production facilities.

Applying design or production rules

Electronics has used logic gate truth tables for years to represent structured 'decision-making' processes. Some electronic systems can be represented graphically using logic gate symbols. Each type of logic gate will respond to specific inputs by generating a predictable output. The inputs and outputs are represented as a '1' or '0', which can be thought of as 'signal' or 'no signal'. A computer printer, for example, can be designed to stop printing if it has either run out of paper or run out of ink. If either or both the ink sensor and the paper sensor send a signal, the printer will stop printing and indicate a problem. This system

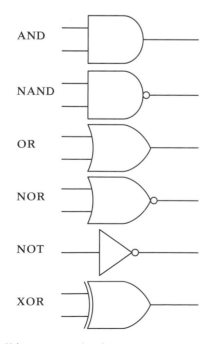

AND

NAND

OR

NOR

NOT

XOR

Fig. 4.11 Logic gates used to design systems

can be represented by a logic gate called an 'OR' gate, which will send an output signal if it receives a signal from either or both of the inputs.

Developing artificial intelligence

Machine vision systems

Machine vision systems are a good example of developing technology. AI systems are combined with cameras and other optical sensors to:

- analyse visual images on the production line for quality, safety and process control
- run product distribution and bar coding systems in computerised warehouses, which can make electronic links between suppliers and customers.

These intelligent 'vision' systems can 'see', 'make decisions', then 'communicate' those findings to other 'smart' factory devices, all in a fraction of a second. Systems have been developed to identify criminal suspects through high street CCTV systems.

Neural networks

Rather than relying on binary systems, neural networks seek to imitate the thought processes of the human brain (which is made up from interconnected neurones). Although computers are very good at performing sequential tasks, the human brain is much better at performing pattern-based tasks that require parallel processing, such as identifying individual voices in a crowd. Neural networks can predict events, when they have a large database of examples to draw on, and are used for voice recognition and natural language processing (NLP).

Voice recognition systems

Voice recognition systems technology will recognise the spoken word but do not understand it. It is used as an alternative computer input and can prove useful in hostile environments, such as space, or when the use of a keyboard is impracticable because the operator is disabled. In the future, an operator will be able to talk directly to an expert system for guidance or instruction.

Natural language processing (NLP)

Communication with computers normally requires us to learn specialised languages. It is hoped that NLP will enable computers to understand human languages. Some rudimentary translation systems that translate from one human language to another are in existence, but they are not nearly as good as human translators.

The use of diagrams to represent simple and complex production systems

Systems and graphical system diagrams

- There are natural systems as well as man-made systems.
- Systems have limits.
- Systems can be broken down into sub-systems.
- Systems are better understood when represented by symbols and diagrams that describe the flow of information and sequence of actions within a process.

Open- and closed-loop control systems

- A system operating **open-loop control** has no feedback information on the state of the output.
- A system operating **closed-loop control** can have either positive or negative feedback.

Positive and negative feedback

Positive feedback increases to match increasing outputs, while negative feedback reduces inputs to a system to avoid instability. Kanbans are

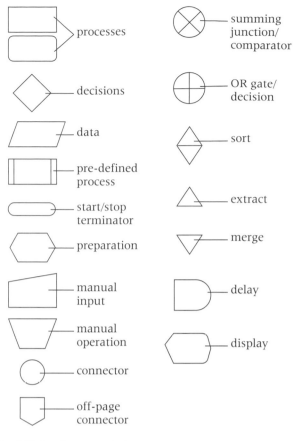

Fig. 4.12 Set of systems symbols

used in some manufacturing systems to control the flow of work on a production line.

Error signals

The difference between the input signal and the feedback signal is called the error signal. If this is higher than required, it will generate a positive error signal to decrease production levels. If production falls, this generates a negative error and production is increased.

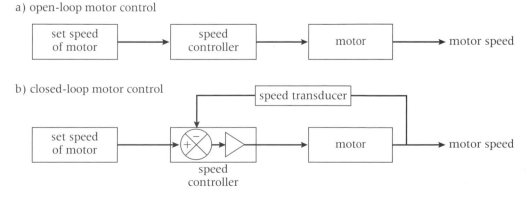

Fig. 4.13 Open- and closed-loop control of a motor

Lag

The time delay before the system is able to respond to failures is known as lag and it is a common feature in closed-loop control systems. Electronic data interchange (**EDI**) and improved 'real time' sales data from electronic point of sale (EPOS) information systems are used to inform the manufacturer of the need to adjust production to correct the 'fault'.

Automated systems using closed-loop control systems

When using a closed-loop system to operate a conveyor belt, feedback is provided by a transducer. This senses when the actual speed of the motor differs from the required speed and sends an error signal to the speed controller. The speed controller will compensate by adjusting the speed of the motor to prevent it overheating.

Sequential control

Robotic and automated processes often use sequential control programs in which a series of actions take place one after another.

Logical control of automated and robotic systems

Combinational logic or 'multiple variable' control requires a series of conditions to be met before an operation can take place.

Fuzzy logic

Fuzzy logic is based upon the observation that good decisions can be made on the basis of non-precise and non-numerical information. Conventionally, computer systems operate on the basis of limited true or false conditions. Fuzzy logic systems can process vague or imprecise information into degrees of truthfulness or falsehood, for example '80 per cent true'. Fuzzy logic has proved to be particularly useful in expert systems, artificial intelligence applications, and database retrieval and engineering.

Fuzzy control

Fuzzy logic controllers (FLC) are the counterparts to conventional logic controllers, such as PLCs. Expert knowledge can be expressed as 'fuzzy sets' in a very natural way using linguistic variables, such as few, very, more or less, small, medium, extremely and almost all. A fuzzy set is a collection of objects or entities without clear boundaries.

Fuzzy control is useful when processes are very complex and non-linear, there is no simple mathematical model, or if the processing of (linguistically formulated) expert knowledge is to be performed.

Applications include:

- PCB design and manufacture
- product analysis
- sampling or testing
- e-commerce applications, such as document management systems, data warehousing and marketing
- control systems in products, such as washing machines, which adjust the washing cycle for load size, fabric type, level of soiling.

The advantages and disadvantages of automation

The pressure for automated manufacturing

Manufacturers are under increasing pressure to devise and develop automated operations that are capable of performing consistently under continually changing or disturbed conditions on a global scale. This is a result of the demand for greater product variety, smaller batch sizes, frequent new product introductions and tighter delivery requirements. In addition,

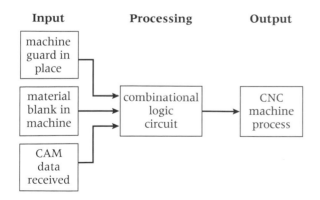

Fig. 4.14 Sensors and software need to be able to verify all three inputs before the CNC machine will operate

Table 4.14 The advantages and disadvantages of automated manufacturing systems

Advantages	Disadvantages
• Reduced labour costs (including compensation costs arising from injury). • Shorter payback time on the capital compared with machines with human operators. • Precision and high speed improves production rates (which typically vary less than three per cent), cycle time, reliability and reduces down time. • Design changes are easier to incorporate. • Faster time to market. • Tooling costs are reduced because complicated jigs and fixtures are not required. • Short set-up time leads to less time required to change from product to product. • Machine tool up time productivity improved by as much as 30 per cent by eliminating production problems, such as bottlenecks. • Sensing, motion, process and system options allow for greater control, consistency and quality output in less time with less chance of scrap or damaged parts. • There is no indirect labour training of potentially large numbers of operators.	Automated systems are not always the most suitable solution and a human workforce can be more cost effective. Too complex a manufacturing process slows down a robot's speed of action and therefore increases manufacturing time. Other significant cost factors include: • the high cost of buying, installing and commissioning • the cost of recruiting and training operators • the cost of keeping up with new technological advances.

manufacturers have to be able to respond to external or internal disturbances to a production process, such as:

• sudden changes in demand for the product
• variations in raw material supply
• machine breakdowns.

The impact of automation on employment

Although automation replaces traditional jobs, there is a shortage of workers equipped with the qualifications and skills to work in modern manufacturing. Automated industries require new attitudes and skills including:

• a wider range of basic skills including literacy and numeracy
• ability to transfer their skills and knowledge in response to the rapid rate of change
• willingness to move jobs
• capability of multi-tasking so that they are competent in responding to different machines within a production cell.

EXAMINATION QUESTIONS

Example questions and answers.

 *a) Name **two** of the four components of the robotic arm illustrated opposite and outline their purpose/function.* **(4 marks)**

Acceptable answer – two of the following

a) **Drive/actuators**: components, such as servo motors or stepper motors, are **used to control movement** in the arm.

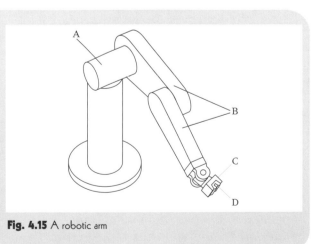

Fig. 4.15 A robotic arm

b) **Link arm(s)**: the parts of the robot that **position the end-effector and sensors**.

c) **End-effector**: the **tool** at the end of the arm, which **performs the programmed task**.

d) **Sensors**: devices that **provide feedback** to the controller, allowing the robotic arm to **monitor or react** to changes in the working environment or manufacturing process.

b) *Industrial robots are very expensive. Describe the advantages of using robots in manufacturing that justify these costs.* **(4 marks)**

Acceptable answer

Robots are used in manufacturing environments for a number of reasons. Most robots can **switch** **effortlessly** between different pre-programmed operations **without the need to stop production**, which reduces lost production in down time. These pre-programmed operations can be **stored indefinitely** until required and **identical components can be produced at any time**. The **consistent accuracy and reliability** of robots using automatic sensors virtually **eliminates human error**. The ability of robots to carry out **hazardous operations** combined with **'fail-safe' procedures** leads to a **safe working environment**. Although robots require a heavy initial investment, these automated systems have much **lower operational costs** and will eventually pay for themselves.

Uses of ICT in the manufacture of products

You need to

understand the impact and advantages/disadvantages of ICT within the total manufacturing process including:

- ☐ electronic communications
- ☐ electronic information handling
- ☐ automated stock control
- ☐ production scheduling and production logistics
- ☐ flexible manufacturing systems
- ☐ production control
- ☐ product marketing, distribution and retailing.

KEY TERMS

Check you understand these terms

email, Internet, EDI, ISDN, LANs, WANs, video conferencing, CAMA, QRM, EPOS, Internet market

Further information can be found in *Advanced Design and Technology for Edexcel, Product Design: Resistant Materials Technology*, **Unit 4B2, section 3.**

KEY POINTS

Electronic communications

Email and modems

Electronic mail or **email** is the simplest form of electronic communication. It has a comparatively low level of reach (level of communication) and range (types of data transfer) when it is used for messaging or sending files to an individual or a work group. Email can be used over intranets, extranets or the **Internet**. The modem allows computers to communicate by converting their digital information into an analogue signal to travel through the public telephone network at a maximum rate of 56 kilobytes per second (kb/s). Advantages include savings in stationery and telephone costs, rapid transmission, all transmissions are recorded, and facilitates work from remote locations.

Electronic data interchange (EDI)

EDI allows users to exchange business documents, such as invoices, delivery notes, orders and receipts, in a similar way to email. ICT is the tool that integrates computer systems and electronic links to create paperless trading. EDI is an essential tool in quick response and JIT systems.

Electronic data exchange (EDE)

CAD/CAM data interchange (CDI) is the process of exchanging design and manufacturing data. The system by which EDI and CDI are combined to provide automated transfer of data over a computer network is called electronic data exchange (EDE). There are various networks available for implementing EDE systems. The key to their usefulness in the field of graphics is their connection speed and the rate at which data can be transferred (throughput).

Integrated services data network (ISDN)

ISDN technology allows multiple digital channels to operate simultaneously through dedicated telephone lines. Advantages include improved data transfer speeds, improved connection speeds, and it dispenses with the need for a modem. The ISDN basic rate interface (BRI) provides two channels of 64 kb/s each or a total of 128 kb/s, and is intended for home-based users. The ISDN primary rate interface (PRI) provides 30 channels of 64 kb/s each or a total of 1920 kb/s, and is intended for business users. ISDN PRI can act as a gateway offering telephone services to users on LAN (local area networks) or can be used to accept large files, which are then distributed.

Broadband

Broadband technology is not available everywhere but is growing in popularity and shares many of the benefits of ISDN. These advantages include:

- up to ten times faster than modem connections
- less expensive than ISDN
- easier to set up (uses existing telecommunications cables) and maintain than ISDN
- more effective support for multimedia and e-marketing features.

Local area networks (LAN)

As the name suggests, **LAN**s are closed networks that are limited to sharing data, information, communications and resources within an organisation.

Wide area networks (WAN)

WANs allow data to be transferred globally using existing digital telephone systems. To ensure compatibility, WANs require dedicated equipment, which can make them expensive.

Intranets and extranets

Intranets use web-based technology to set up local networks that can be password protected. Web browsers are used to navigate HTML pages and 'firewalls' protect the network from unauthorised, external access.

An extranet is used to share data with business partners and customers using the Internet. Levels of access to sensitive areas of the extranet are protected by passwords. Subscription services use this technology to protect 'expert knowledge' services, for example.

Global networks (the Internet)

The Internet is the international computer network linking together thousands of individual networks. The world wide web (www) is the familiar collection of inter-connected documents and files, such as websites, which are accessible through the Internet. These websites are held on Internet servers that process and communicate data via cable, radio and satellite. Features of the Internet and world wide web include:

- ISPs (Internet service providers), companies that provide access to the Internet
- web browsers, such as Netscape Navigator and MS Explorer, used to navigate the Internet
- search engines, such as Google, Alta Vista, Ask Jeeves, used to find information sources
- URL (universal resource locator), the unique address allocated to each website
- HTML (hypertext mark-up language), the language used to write web pages
- hyperlinks, the 'hotspots', such as buttons or underlined text, used to move between pages.

Advantages of using the Internet and the web include:

- a low-cost, easily accessible means of sharing ideas within interest groups
- an almost infinite source of information (which may or may not be useful or accurate)
- a medium for communicating with current and potential customers
- a means of researching what other designers or manufacturers are producing
- a readily accessible online reference source of commercial data.

Disadvantages of using the Internet and the web include:

- industrial espionage and 'hackers'
- computer viruses
- fraud.

Video conferencing (VC)

Video conferencing allows individuals in remote locations to hold 'virtual meetings'. VC is available in two forms.

1 Desktop video conferencing (DTVC) designed for home user two-way communication.
2 Multi-point video conferencing for business users who are able to conduct 'virtual conferences'.

VC has developed as a result of advances in processing power, electronic communications (including ISDN and broadband), and digital video technologies that allow people in different parts of the world to hold virtual meetings.

The advantages of video conferencing

- On some VC systems, data, such as CAD drawings, can be transferred during the meeting.
- It removes the need to travel to meetings, such as marketing presentations, saving time, travel expenses and stress.
- The ease of use enables designers, manufacturers and executives to meet, ensuring regular communication, immediate decisions and close control of the development process.
- Education and training can be carried out more efficiently. Expertise can be shared across the company without the need to travel.
- Problems can be solved much more effectively (remote diagnostics) since the relevant experts can meet immediately to address the problem, reducing lost production time.

Remote manufacturing

The ability to communicate instantly, using video conferencing, with people in any continent combined with new technologies allowing reliable electronic data exchange means that designs can be manufactured anywhere in the world.

- Video conferencing and other communication technologies ensure that designs can be developed in discussion with the manufacturer to ensure they meet the constraints of the manufacturing process.
- The finished design can be sent electronically, directly to the manufacturing centre where it is machined using CNC equipment. The process can be monitored by the design team using video.
- The whole process is very quick and finished components can be checked and dispatched the same day.
- The whole process allows designers the opportunity to take advantage of CNC manufacturing technology without having to make the heavy investment necessary to purchase all the machinery.

New communications technology

Electronic whiteboards (interactive or smart boards)

An electronic whiteboard can be used in presentations, video conferences, and training sessions and for recording data. Electronic whiteboards can provide the following features:

- an interactive writing surface
- a scanner and thermal printer for producing hard copy
- access to computer software, data and video images
- automatic recording functions
- remote control devices, such as wireless touch sensitive tablets.

Information centres or PC kiosks

Interactive kiosks process, communicate and display graphic information and data stored on a computer or network. They can be accessed 24 hours a day from a touch-screen, keyboard or mouse-driven interface and provide services including:

- multimedia presentations of local information for the tourist trade
- video teleconferencing and public Internet access
- interactive services at museums, galleries and trade and product shows.

Electronic information handling

Features of agile manufacturing

Features of agile manufacturing include:

- flexible manufacturing systems (FMS) and quick response manufacturing (QRM)
- customer-driven rather than production-driven manufacture
- emphasis on quality
- close partnerships with customers and suppliers
- information rich
- ICT centred.

Computer-aided market analysis (CAMA)

Market research helps companies to predict demand, identify potential markets, target specific market groups (niche markets), identify market trends and tailor marketing strategies. **CAMA** is an ICT-driven strategy through which data is gathered from sources, including surveys and questionnaires, by manufacturers or specialist agencies. Relational databases are used to store, analyse and present this data as useful information.

- A qualitative analysis will provide information on customers and their opinions.
- A quantitative analysis will provide facts and figures, such as sales figures or financial information.
- Regional trend analysis will provide performance information on a geographical basis.

- Market timing attempts to predict future market trends, which will help investment decisions.
- Customer profiling using existing customers can help to identify future markets.

Benefits of CAMA include:
- large amounts of marketing data can be processed into useful information very quickly
- demand and trends can be calculated accurately leading to formation of sales targets and marketing strategies
- direct marketing initiatives, such as product launches, can be directed at specific target markets very precisely (market segmentation).

Computer-aided specification development

Complex products need complex specifications that take account of aesthetic requirements, functional requirements, ease of manufacture (design for manufacture – DFM), and ease of assembly (design for assembly – DFA).

Integrated ICT-based systems already exist where design features can be generated by CAD software and checked by a knowledge-based expert system for ease of manufacture and assembly.

Design specifications are generated with the help of computer technology. These specifications should contain all the information necessary to allow manufacturing to take place. Details can be drawn from a product data management (PDM) system. There are three classifications of information or knowledge held within an intelligent design system.

1 CAD data contains specific information about the physical characteristics of each component part being designed.
2 The design catalogue is a reference for data, such as the costs and properties of standard materials and components.
3 The knowledge database contains 'rules' about design and manufacturing methods.

Automated stock control

Manufacturing systems such as JIT rely on sophisticated, automated and ICT-centred stock control systems. Bar codes or other methods of identification allow materials and components to be monitored in real time throughout the process and waste is minimised. Automatic storage and retrieval systems (ASRS) and automated guided vehicles (AGVs) can be used for materials handling.

- Waste is minimised.
- The inventory is optimised but available on demand.
- Automated stock control systems enable the move from batch to continuous flow production.
- Waiting times caused by unbalanced production times are reduced through the use of scheduling techniques, such as line balancing.

Production scheduling and production logistics

Because manufacturing systems, such as FMS, involve major capital investment, it is important to get the most out of the equipment. Traditional scheduling is generally quite rigid and production will stop in response to changes in production or mechanical breakdown. Computer-based scheduling and logistics systems ensure that production is 'smoothed' so that small variations in supply and demand are managed without causing problems. This is achieved by careful planning, which spreads the product mix and the product quantities evenly over each day in a month. The advantages of computer-based production scheduling include:

- flexibility and responsiveness to changing conditions
- optimisation of work in progress
- reduction in the inventory
- production is balanced between the stations on the production line
- increase in productivity levels.

Flexible manufacturing systems and QRM

Quick response manufacturing (QRM)

Features of **QRM** include:

- reducing product lead times

- rapid production of small or large batches
- ability to automatically reprogramme manufacturing and business processes in response to market pressures
- stock levels are constantly evaluated and adjusted to reflect changing demand patterns
- individuals are able to review changes globally and are automatically alerted to changes
- information flow is carefully managed by PDM systems using highly integrated knowledge bases.

Production control

Quality monitoring (quality control)

Automated inspection systems gather and analyse data in order to provide feedback, which is used to make automatic adjustments to the manufacturing process. Automated quality monitoring systems rely on a large range of sensor technologies including:

- mechanical methods, which use probes and sensors to collect physical data, such as dimensions
- optical quality monitoring systems, which use scanning technology, optical devices, digital cameras and vision systems along with sensors to collect optical data, which is automatically compared with the specification tolerances.

Using digital cameras for monitoring quality

Digital cameras connected to a computer or to a dedicated microprocessor allow online inspection. Advantages include:

- 100 per cent piece-by-piece inspection
- realtime quality control and fast response time
- collected data is automatically compared with specifications, evaluated and stored
- audible or visual signals can be used to alert operators to problems as they arise
- systems can incorporate automated responses
- no direct mechanical contact with the product
- cameras can be placed well away from machinery.

Product marketing, distribution and retailing

Electronic point of sale (EPOS)

EPOS systems use barcodes and laser-operated readers, which generate a large amount of information, to keep track of products throughout the supply chain. The software allows a two-way flow of data and information. When a product is sold at the checkout in the supermarket, the fact is recorded and used to order replacement goods.

Advantages include:

- company financial performance can be monitored at all times due to the availability of detailed and up-to-the-minute records of transactions
- detailed sales histories can be used to predict future trends and fluctuations in demand
- companies can react quickly to changes in demand because the system will inform them instantly of unpredicted changes in consumer-buying patterns
- distribution chains can trace the progress of deliveries to ensure efficiency and allow transferral of products within the company
- real time stock updates allow suppliers and retailers to maintain minimal stock levels, saving money and resources; daily deliveries support continual product replenishment (CPR)

- allows a two-way flow of financial data, emails, price updates and information.

Internet marketing (e-commerce)

Many companies use the Internet, not only to promote their products and services, but as a means of promoting and selling their products and services (**Internet marketing**). In order to succeed, companies need to restructure their internal and external business processes, supply chains and relationships. Some companies trade exclusively on the Internet, taking advantage of:

- cost-effective access to a global market place for relatively small companies and large organisations alike, increasing the customer base and company profile
- all business can be conducted from one geographical site, reducing costs
- a large part of the transaction process can be automated, resulting in faster processing of orders and transactions, reducing overheads and the need for sales staff
- enables the collection of detailed customer profiles
- reduces time to market
- product information is easily accessible and can be changed or updated easily. Virtual products provide access to detailed product information
- use of integrated ICT systems throughout the process leads to a very fast and efficient process.

EXAMINATION QUESTIONS

Example questions and answers.

a) *The Internet is a widely used resource. Explain the following terms with reference to the Internet:*

- HTML
- hyperlinks
- web browsers. **(3 marks)**

Acceptable answer
HTML: hypertext mark-up language is the language **used to create web pages**.

Hyperlinks: hyperlinks are the **underlined words or symbols** that enable you to **jump from one web page to another**.

Web browsers: Microsoft Explorer is an example of a web browser **program** that allows people to **navigate the Internet**.

b) *Discuss the advantages and disadvantages of using the Internet and world wide web for designers developing a new product.* **(8 marks)**

Acceptable answer

The Internet and world wide web provide the product designer with a valuable resource. When developing a new graphic product, the Internet provides designers with a **low-cost, easily accessible means of sharing design ideas** with other colleagues and professionals. The Internet contains a vast amount of copyright-free **information and design images**, which may be incorporated into designs, but users must be aware that this information **may or may not be useful or accurate**. As the designer develops his or her ideas, the Internet provides a very useful means of **communicating with the client** using email or video conferencing. Commercial websites can provide **valuable information about competing products**. When new products are launched, the Internet can be used as one **platform to promote** the new product by creating a website and through the use of e-marketing. However, connecting to the Internet opens the company to hostile attack from **hackers who may be looking for commercially sensitive information** on the company network. It also leaves the company open to attack from computer **viruses**, which are often spread by email and which can cause **permanent damage to important data**.

PRACTICE EXAMINATION STYLE QUESTION

1 a) Outline the aims, features and effects of just in time (JIT) manufacturing. **(5)**

 b) Describe how automated control and feedback systems are used in modern industry. Refer to **all** of the following in your answer:
 • materials handling
 • materials processing
 • quality control
 • safety systems
 • co-ordination of production. **(5)**

 c) Artificial intelligence is a rapidly developing technology.
 i) Explain the term 'artificial intelligence'. **(1)**
 ii) Explain **two** of the following terms in relation to AI:
 • expert systems
 • computer/machine vision
 • natural language processing (NLP)
 • artificial neural networks
 • fuzzy logic. **(4)**

2 a) Referring to an industry of your choice, describe how computer integrated manufacture (CIM) is used to enhance the manufacturing process. **(5)**

 b) Discuss the advantages and disadvantages associated with automated manufacturing systems. **(5)**

 c) Describe the benefits of video conferencing technology compared with more conventional methods of communication. **(5)**

Total for this question paper: 30 marks

---

4 B3 Mechanisms, energy and electronics (R404)

This option will be assessed in section B during the $1\frac{1}{2}$ hour, Unit 4 examination. If you have chosen this option you should spend half of your time (45 minutes) answering all the questions in this section. It is important to use specialist and technical language in the exam, along with accurate spelling, punctuation and grammar. Where appropriate you should also use clear, annotated sketches to explain your answer. *You do not have to study this chapter if you are taking the Design and Technology in Society option, or the CAD/CAM option.*

Pulleys and sprockets

You need to

☐ **be able to select appropriate types of transmission systems for specific applications**
☐ **understand how tension is maintained in drive belts.**

KEY TERMS
Check you understand this term

jockey wheel

Further information can be found in *Advanced Design and Technology for Edexcel, Product Design: Resistant Materials Technology*, Unit 4B3, section 1.

KEY POINTS

Selecting transmission systems

Belts and pulleys and *chains and sprockets* are both types of transmission systems. Both transmit rotation from one shaft to another by means of a flexible belt running on a pulley or by a chain running on a sprocket. Pulleys are available in a number of forms, which have specific types of belt that have to be used with them (Table 4.15 overleaf).

The major disadvantage with any pulley system is its tendency to slip. Chains and sprockets, such as those found on a bicycle or motorbike, use a positive means of location and therefore make it a much more efficient method of transmitting motion. The drawbacks, however, are that they require much more maintenance, especially oiling.

Maintaining tension in drive systems

In an attempt to make pulleys and belts as efficient as possible, every effort is made to maintain the tension in the belt. If a belt is too loose, it may slip; if it is too tight, it might cause the shafts to bend. Tension can be maintained by having adjustable pulleys that can, once the belt has been put in place, be stretched and tightened. Another way is to use a **jockey wheel**. This is a small pulley that is held under tension against the pulley with a spring. This system keeps the belt running smoothly whilst maintaining the correct level of tension.

Table 4.15 Belt types

Belt pulley type	Applications
Flat	The belts are made from neoprene rubber with nylon inserts. Their flat shape allows them to be twisted, meaning that they can be used to transmit motion through 90 degrees. They can also be crossed over to give a change in rotational direction. This type of belt is a general purpose belt and used in basic transmission systems.
V	V belts are much less likely to slip because of their shape, which tends to pull them in. They are commonly used for electric motor drives and car fanbelts.
Toothed	Toothed belts have a series of teeth moulded on the inside. They are used when it is critical for the driving pulley to move in sequence with the driven pulley. They are used on X-Y plotters and for car timing belts.
Coned	This type of belt can adjust its speed and position whilst still rotating. The speed is controlled by moving the pulley position. In some instances, the belts have a series of radial teeth on the inside. They are used in machines that need variable speeds, such as wood-turning lathes.

EXAMINATION QUESTION

Example question and answer.

Q1 *Give **one** disadvantage of using a chain and sprocket rather than a pulley and belt.* **(2 marks)**

Acceptable answer

A chain and sprocket needs much **more regular servicing** than a pulley and belt. This is because there are **many links in a chain and each one needs to be able to rotate freely, so it must be cleaned and oiled regularly.**

Shafts and couplings

You need to

☐ **understand how aligned and non-aligned shafts can be coupled**
☐ **know about flexible and universal joints**
☐ **know how wheels and couplings can be fixed to shafts.**

KEY TERMS

Check you understand this term

coupling

Further information can be found in *Advanced Design and Technology for Edexcel, Product Design: Resistant Materials Technology*, Unit 4B3, section 2.

KEY POINTS

Coupling of aligned and non-aligned shafts

A **coupling** is a joint that is used to connect shafts together while still enabling rotary motion and torque to be transmitted. Couplings can be made in a number of ways:

- solid bolted coupling
- muff coupling
- compression.

These methods can only be used to connect aligned shafts, those which share the same centre line, and all involve bolting stiff sleeves or collars around the ends of the two shafts.

Flexible joints

Flexible joints are used where the shafts are not in perfect alignment. They are also used to absorb small vibrations and shocks. The joints are normally formed from a disc of rubberised material sandwiched between two metal discs or 'spiders'. Although the elastic nature of the material absorbs the shock and vibration, it does have a tendency to harden and perish over time.

Universal joints

Where shafts are more than a few degrees out of alignment, a universal joint is used. Non-aligned shafts of up to 20 degrees can be connected using universal joints, the most common type being a Hooke's universal joint. The major problem with universal joints is that the output shaft does not run at a constant velocity. To overcome this, two joints have to be used, the second joint being at the same angle as the first joint. For angles greater than 20 degrees, a constant velocity joint must be used.

Fixing wheels and couplings to shafts

All pulleys and couplings need to be fixed. It is essential that they are all fitted securely (Table 4.16).

Table 4.16 Fixing to shafts

Method of fixing	Description
Keys and keyways	Keys and keyways provide a secure and positive fixing method. A keyway is a groove that is cut into the shaft on a milling machine. The key is a hardened tapered wedge. The pulley is often also cut with a groove and when the pulley is correctly aligned on the shaft, the key is driven in to hold the two parts together.
Cotter pins	These are commonly used to hold bicycle pedals onto the crankshaft. A cotter pin is a tapered pin and is either rectangular or circular in cross section. The cotter pin is tapped into the hole and then held firmly in place by a nut on the underside.
Splined shafts	Splined shafts are used where pulleys need to be able to move longitudinally along a shaft whilst still rotating. The splines, or series of grooves in the shaft, stop the pulley from rotating.

Table 4.16 Fixing to shafts *(continued)*

Method of fixing	Description
Grub screws 	Grub screws are generally used where little force is involved. A hole is drilled in the pulley flange and is threaded. A flat is fitted on the shaft and a small grub screw holds the pulley onto the shaft. In some cases, a second screw is inserted to stop the first one becoming loose from vibration.

EXAMINATION QUESTION

Example question and answer.

Q1 *Shafts can be coupled in many ways, such as:*

- *flange couplings*
- *flexible joints*
- *universal joints.*

*Select **one** of the above and use brief notes to explain how it works.* **(2 marks)**

Acceptable answer

Flexible joints are used when the shafts are not in perfect alignment, but by up to only a few degrees. **They are also** used **where small vibrations are encountered.** They make use of a **small rubber disc trapped in between two metal plates, sometimes referred to as 'spiders'.** As the shafts rotate, any shock is absorbed by the rubber plate, and if the shafts are not perfectly aligned, the rubber is either compressed or stretched.

Cams and followers

You need to

recognise and understand:

- ☐ the uses of cams
- ☐ the types of followers.

 KEY TERMS
Check you understand these terms

cam, rise, fall, dwell, follower

📖 **Further information can be found in** *Advanced Design and Technology for Edexcel, Product Design: Resistant Materials Technology,* Unit 4B3, section 3.

KEY POINTS

The uses of cams

A **cam** is a mechanism used to change an input rotary or oscillating motion into either a reciprocating or oscillating output motion (Table 4.17). Cams must be securely fixed to a rotating shaft using any of the methods previously described.

Types of followers

A **follower** is a device that is held against the cam profile and follows the shape or profile of the cam. Followers are available in a range of profiles and must be carefully chosen to match and fit the profile of the cam being used. Followers can be flat, knife-edged, points or rollers (Table 4.18).

Table 4.17 Cams and their applications

Cam type	Description/application
Pear-shaped	This type of cam, as its name suggests, is pear-shaped. They are used to control the opening and closing of inlet and exhaust valves in engines. They are symmetrical in shape and so the **rise** and **fall** times are identical, although this period is governed by the stroke of the cam, the distance between the tip of the cam and the **dwell** or rest spot.
Eccentric	This is the simplest form of cam since it is an off-centre circular disc. This cam gives a smooth rise and fall, giving rise to simple harmonic motion. The output motion can also be described as constant acceleration and retardation. Eccentric cams are used in fuel pumps.
Heart-shaped	Heart-shaped cams are symmetrical and generate a continuous motion. They are used as bobbin winders to wind relay coil and motor windings.
Uniform acceleration/ retardation cams	This type of cam generates an output with uniform acceleration and retardation.

Table 4.18 Cam followers

Follower	Description
Flat	A flat follower generates a great deal of friction because of the surface area in contact with the cam. It cannot be used to follow hollow cams and contours.
Knife-edge	A knife-edge cannot be used to follow hollow contours because it cannot generally lift out of the cam. It does, however, give the most accurate conversion of movement.
Point	A point gives an accurate transfer of movement from the cam but the point and cam undergo rapid wear. It can be used on hollow cams.
Roller	Although this is the most expensive type of follower, it causes the least wear on the cam. It cannot, however, be used to follow hollow cam profiles.

EXAMINATION QUESTION

Example question and answer.

Q1 *Cams are available with numerous profiles. Select **one** form of profile and briefly explain how it can be used.* **(2 marks)**

Acceptable answer

A **heart-shaped cam is symmetrically shaped and produces a constant acceleration and retardation**. This makes it **ideal for winding wire and thread onto drums and bobbins**.

Conview of motion

Conversion of motion

You need to

☐ understand how rotary motion can be converted in linear motion and vice versa.

> **Further information can be found in** *Advanced Design and Technology for Edexcel, Product Design: Resistant Materials Technology*, **Unit 4B3, section 4.**

KEY TERMS

Check you understand these terms

rotary, linear, reciprocation

KEY POINTS

The conversion of motion

Crankshafts, cranks and sliders, rack and pinion, and cams are all mechanisms that can be used to convert rotary motion (Table 4.19). A pawl

Table 4.19 Methods of converting motion

Mechanism type	Description
Crank and slider 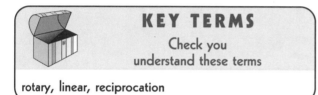	This mechanism converts rotary into reciprocating motion. As the crank rotates, the slider reciprocates. The longer the crank, the further the slider moves. It is possible to use the mechanism in reverse to convert reciprocating motion into rotary motion. Crank and sliders are used in engines where the pistons are connected to a crankshaft.
Rack and pinion Pinion Rack	Rack and pinions are used to convert rotary motion into linear motion or vice versa. A spur gear, known as the pinion, meshes with a rack, which has teeth set in a straight line. If the pinion rotates, the rack will move in a straight line; if the rack is made to move, the pinion will rotate. This type of mechanism is used on canal lock gates where they raise and lower the sluices. They are also used on pillar drilling machines in workshops where they make the rotating chuck move down into the work being drilled.

Table 4.19 Methods of converting motion *(continued)*

Mechanism type	Description
Cam and follower	As the cam is made to rotate on the shaft, the follower follows the cam profile. The follower in this instance is made to reciprocate. Cams and followers like this are used to control the opening and closing of inlet and exhaust valves in car engines.
Pawl and ratchet	A pawl and ratchet is used in lifting mechanisms where rotary motion winds a cable onto a drum and the load is lifted. In this instance, when heavy loads are involved, a pawl and ratchet is used to prevent the shaft from unwinding and dropping the load. The pawl is spring loaded, which holds it firmly in place against the ratchet wheel. When the shaft is needed to rotate in the opposite direction, the pawl is released and the ratchet is then free to rotate.

and ratchet is used to arrest or stop rotary motion. **Rotary** motion is a circular type of motion, like the movement of hands on a clock. **Linear** motion is movement in one direction in a straight line. **Reciprocation** is when an object moves backwards and forwards or up and down in a straight line, like a saw being used to cut a straight line in a piece of wood.

EXAMINATION QUESTION

Example question and answer.

Q1 *Rotary motion can be converted into linear or reciprocating motion in many ways. Select **one** method and, using notes and sketches, explain how the conversion process takes place.* **(2 marks)**

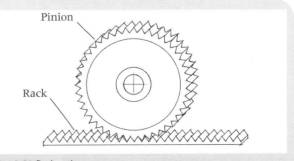

Fig. 4.16 Rack and pinion

Acceptable answer
A **rack and pinion** is one method of converting rotary to linear or reciprocating motion. This process can also be used in reverse to convert linear or reciprocating motion into rotary motion. **A spur wheel, known as a pinion ,meshes with a rack. As the pinion rotates and the teeth mesh, the rack moves.**

Brakes

You need to

☐ **recognise types of brakes.**

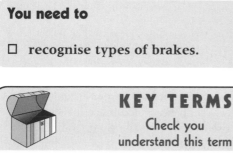

KEY TERMS
Check you understand this term

friction

📖 **Further information can be found in** *Advanced Design and Technology for Edexcel, Product Design: Resistant Materials Technology*, **Unit 4B3, section 5.**

KEY POINTS

Types of brakes

All forms of brakes are mechanical systems (Table 4.20). Any moving body has kinetic energy, and braking systems are used to absorb this energy in order to bring the moving body to rest. **Friction** between the brake linings and the wheels that grip the road, slows the rotational speed and makes the vehicle slow down and eventually stop.

Table 4.20 Brake types

Brake type	Description
Cantilever brakes	Bicycle brakes are operated by pulling a lever. This pulls a brake cable that sits inside a sleeve and is attached to a metal bracket at the other end. The tension put on the cable pulls the two swing arms together and in turn brings the two rubber brake blocks to bear on the metal wheel rim. Once the bike has stopped and the lever released, the spring mechanism opens the arms, releasing the brake blocks from the wheel rim.
Disc brakes	Disc brakes are commonly used on the front wheels of most cars because they have a greater braking capacity than drum brakes, and since the weight of the car is transferred forward on braking, the most efficient method must be used. They consist of a metal disc that is attached to the hub or axle. The pressure is applied via a brake calliper activated by the pressing of the brake pedal inside the car. Since it is a hydraulic system, fluid is pressurised and is quite capable of exerting great forces.
Drum brakes	Drum brakes are used on the back wheels of cars because they take less of the braking force. They can be operated both hydraulically and mechanically and therefore the handbrake is connected to them. Two shoes are placed inside a rotating wheel hub; each shoe has an asbestos-based friction lining. When the brake pedal or handbrake is applied, the two shoes are forced out to rub against the hub in order to slow the wheel down.

Clutches

You need to

☐ recognise and describe the action of a clutch.

KEY TERMS

Check you understand these terms

cone, single plate, diaphragm, centrifugal

Further information can be found in *Advanced Design and Technology for Edexcel, Product Design: Resistant Materials Technology*, Unit 4B3, section 6.

KEY POINTS

The action of a clutch

A clutch is a type of shaft coupling that allows a rotating shaft to be easily connected to or disconnected from a second shaft (Table 4.21). It is essential that the two shafts are aligned since a clutch cannot be used on non-aligned shafts. Clutches can be divided into two categories: positive and friction clutches.

• A positive clutch can only be used if the driving shaft is brought to rest before trying to engage it. The ends of the two shafts have a series of claws fitted to them. As the two sets of claws are brought together, they mesh.

Table 4.21 Clutch types

Clutch type	Description
Cone clutch	The cone clutch, one of the commonest types of clutch, belongs to the friction clutch group. The two separate parts are both cone shaped; as the two parts are brought together, the two parts of the clutch join together to make one.
Single plate clutch	A single plate clutch allows for very quick engagement or disengagement. The pressure is applied via a number of coil springs and the motion is transferred when the friction plate is firmly clamped.
Diaphragm	The diaphragm spring clutch is generally much more compact than a single plate clutch. It is simpler to use and maintain since the springs are replaced by a single diaphragm spring.
Centrifugal	The centrifugal clutch is configured in a similar fashion to a pair of drum brakes. Two shoes are pivoted and retained by springs on a central plate. As the speed of the rotating shaft increases, the centrifugal forces move the shoes out towards the outer drum. When rotating fast enough, the friction between the two surfaces is enough to engage the outside drum. These types of clutches are used on lawn mowers and go-karts.

• Friction clutches are used in situations where it is not always possible to bring the rotating shafts to a stop. Friction clutches rely on the two surfaces being brought together under pressure and the forces between the two being enough to make them rotate together at the same speed.

Examiner's Tip

In your responses, remember that one-word answers will not be sufficient. You must always justify your responses.

EXAMINATION QUESTION

Example question and answer.

Q1 *Explain the difference between a positive and a friction clutch.* **(4 marks)**

Acceptable answer
A **friction clutch relies on two surfaces coming together under pressure**. The frictional forces are such that **the two surfaces both rotate together at the same speed**. **A positive clutch relies on the two shafts being at rest**. Since they have a **series of teeth, they have to be brought together so that the teeth mesh**. If this is attempted with rotating shafts, **the teeth can be badly damaged and knocked off**.

Screwthreads

You need to

☐ **understand how mechanical advantage is gained through the use of an inclined plane and in the screwthreads**
☐ **recognise and describe a use for screwthread forms.**

KEY TERMS

Check you understand these terms

inclined plane, helix

Further information can be found in *Advanced Design and Technology for Edexcel, Product Design: Resistant Materials Technology*, **Unit 4B3, section 7.**

KEY POINTS

Gaining mechanical advantage

An **inclined plane** is simply a sloping surface that can be used to gain a mechanical advantage (MA) when raising a load. It is much easier to raise a heavy load up a slope to the required height than to lift it vertically.

If the inclined plane is wrapped around a cylindrical shaft, a **helix** is formed. A screwthread is then formed if a groove is cut along the helix. A screwthread then imitates the action of an inclined plane and is used to transmit motion and force with the same mechanical advantage as that of an inclined plane (Table 4.22).

Table 4.22 Uses for screwthread forms

Screwthread form	Description
V-thread	The V-thread takes its name from the V-shaped profile of the thread form. A large amount of friction exists with this form; they are used extensively for fastenings, such as nuts and bolts.

Table 4.22 Uses for screwthread forms *(continued)*

Screwthread form	Description
Square	As its name suggests, the profile of the thread form is square. It is generally used in lifting applications, such as car jacks, or for moving parts of machinery. They are also used in sash and G clamps where they are able to exert large forces.
Acme	Acme threads are used on machines where the nut needs to engage onto a rotating shaft. This is primarily used on a centre lathe where there is a lead screw that rotates. By engaging the nut, the cutting tool is made to travel along the work automatically. This thread form is ideal for this because its sloping sides allow the nut to engage easily.
Buttress	A buttress thread is used with quick-release mechanisms but only allows the application of force in one direction.

EXAMINATION QUESTION

Example question and answer.

Q1 *Describe how a screwthread makes use of an inclined plane.* **(2 marks)**

Acceptable answer

An inclined plane is simply a sloping surface. If an **inclined plane is wrapped around a cylinder, a helix is formed, which is the basis for any form of screwthread.**

Transistors

You need to

- ☐ **understand how a Darlington pair is used and configured**
- ☐ **know how to use a transistor as a transducer driver**
- ☐ **know that a Field Effect Transistor (FET) is used to drive high-current devices.**

KEY TERMS

Check you understand this term

transducers

 Further information can be found in *Advanced Design and Technology for Edexcel, Product Design: Resistant Materials Technology*, Unit 4B3, section 8.

KEY POINTS

Darlington pairs

Transistors can be used as electronic switches and current amplifiers. By controlling the voltage across the base emitter junction, the transistor can be turned on and off. This is known as the *threshold voltage*. Transistors are selected on one of two factors, their gain or the maximum amount of current that can pass through the collector. No transistor, however, has a high gain and a high I_c capability. By connecting two transistors together, one with a high gain and one with a high I_c capacity, the two functions can be combined. This arrangement is called a Darlington driver. Since two transistors are now used, the threshold voltage is 1.2V rather than 0.6V for a single transistor. The overall gain for a Darlington driver is also increased and is simply calculated by multiplying the gain of the two single transistors together.

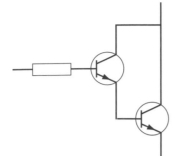

a) Darlington pair

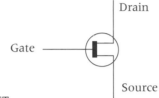

b) FET

Fig. 4.17 Transistors in use

Transistors as transducer drivers

Transducers are output devices, such as bulbs, motors, relays, solenoids and speakers. These devices are sometimes known as the *load*, and they can be turned on and off by the switching action of the transistor. As the load, they are connected in between the collector and the positive supply rail.

Field Effect Transistors (FETs)

Field Effect Transistors are voltage amplifiers rather than current amplifiers. They are sensitive to voltage changes at the gate terminal and can be used to handle currents of 2–12 amps. In normal use, the drain is connected to the positive supply rail via the load, the source is connected to 0V and the gate to the switching input.

EXAMINATION QUESTION

Example question and answer.

Q1 *Explain why the threshold voltage for a Darlington pair is 1.2V.* **(2 marks)**

Acceptable answer

Since a **Darlington pair is made from two transistors, each with a threshold voltage of 0.6V, the combined total is 1.2V.**

Thyristors

You need to

☐ understand how a thyristor is used to latch an output.

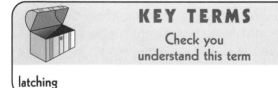

KEY TERMS

Check you understand this term

latching

📖 **Further information can be found in** *Advanced Design and Technology for Edexcel, Product Design: Resistant Materials Technology,* Unit 4B3, section 9.

KEY POINTS

The thyristor as a latch

A thyristor is used where an electronic latch is required. **Latching** means that when the system is turned on, it remains on until reset. When triggered with a small current on the gate, it turns on, and remains on; this action makes it very useful in burglar alarms.

The voltage at the gate controls the flow of current between the anode and cathode in a similar way to that of a transistor. When a positive voltage of approximately 2V is applied to the gate, a large current will flow from the anode to the cathode. This current continues to

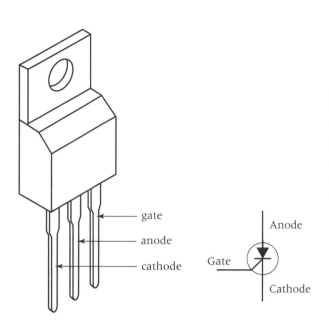

gate
anode
cathode

Anode
Gate
Cathode

Fig. 4.18 The thyristor

flow even if the voltage at the gate is removed or falls below 2V.

Like the transistor, the thyristor must be protected on the gate with a current-limiting resistor since they are current-sensitive devices. The circuit diagram in Figure 4.18 shows how the thyristor should be configured. The push-to-make switch has been connected across the anode and 0V and this is used to reset the thyristor.

Logic

You need to

☐ understand and recognise logic gates and draw truth tables for them
☐ be able to use two input logic gates to solve combinational logic problems
☐ understand that logic gates respond to digital signals
☐ describe digital and analogue signals.

Further information can be found in *Advanced Design and Technology for Edexcel, Product Design: Resistant Materials Technology*, Unit 4B3, section 10.

KEY TERMS

Check you understand these terms

logic gates, digital signals, analogue signals

KEY POINTS

Logic gates and their truth tables

Logic gates exist in many forms, with the NAND gate internationally recognised as the most common. It is possible to make the other types of gates from NAND, partly explaining its popularity (Table 4.23).

Table 4.23 Logic gate and truth table summary

Gate type	Symbol	Truth table
AND	A, B → Q	A B Q / 0 0 0 / 0 1 0 / 1 0 0 / 1 1 1
NAND	A, B → Q	A B Q / 0 0 1 / 0 1 1 / 1 0 1 / 1 1 0
OR	A, B → Q	A B Q / 0 0 0 / 0 1 1 / 1 0 1 / 1 1 1
NOR	A, B → Q	A B Q / 0 0 1 / 0 1 0 / 1 0 0 / 1 1 0
NOT	A → Q	A Q / 1 0 / 0 1
XOR	A, B → Q	A B Q / 0 0 0 / 0 1 1 / 1 0 1 / 1 1 0

Combinational logic problems

Combinational logic involves using more than one type of gate to solve a logic problem. Boolean algebra is used to solve complex logic problems where a set pattern of input conditions have to be met to achieve the desired output. Larger truth tables are used to help map the outputs.

Logic gates responding to digital signals

Logic gates only respond to **digital signals**. A digital signal can either be ON or OFF. These two states can be described in a number of ways (Table 4.24). In most cases, the inputs to logic gates are controlled by switches and therefore provide digital pulses.

Table 4.24 Summary of logic level

ON	HIGH	1
OFF	LOW	0

Digital and analogue signals

Electronic circuits and components can behave in one of two ways: analogue and digital (Table 4.25).

Table 4.25 Analogue and digital signals

Mode of operation	Characteristics
Analogue	**Analogue signals** vary and can exist at any level between the maximum and minimum voltage levels at any one stage. For example, to say it is hot or cold does not really give a true indication of the temperature. The temperature fluctuates between extreme hot and cold. Logic gates cannot cope with analogue signals and they have to be converted into digital signals.
Digital	Digital signals exist in only two states: on or off, high or low, 0 or 1. Temperature fluctuations cannot be monitored unless they are referenced against a certain point: hot or cold. Logic gates and circuits respond to digital signals.

EXAMINATION QUESTION

Example question and answer.

Q1
a) Draw the symbol and complete a truth table for a two-input NAND gate. **(2 marks)**

b) Show how NAND gates can be combined to behave as an OR gate. **(3 marks)**

A	B	Q
0	0	1
0	1	1
1	0	1
1	1	0

Acceptable answer

a)
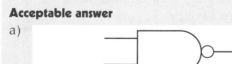
Fig. 4.19 A two-input NAND gate

b)
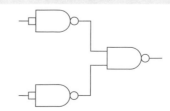
Fig. 4.20 Combined NAND gates

Circuit construction

You need to

☐ be able to construct circuits using temporary and permanent methods of construction and assess the usefulness of each.

Further information can be found in *Advanced Design and Technology for Edexcel, Product Design: Resistant Materials Technology,* **Unit 4B3, section 11.**

KEY POINTS

Constructing circuits and assessing the usefulness of the methods used

When designing and making electronic products, it is important to use modelling techniques to test the circuit that has been designed. This should be done before producing a final printed circuit board (PCB). Many methods of modelling are available, some more permanent than others (Table 4.26).

Table 4.26 Circuit construction techniques

Construction technique	Description
Breadboard/ prototyping board	Breadboards are used for modelling and testing designs before producing PCBs. They are ideally suited for use with ICs (integrated circuits), which fit neatly down the centre of the board. All the individual columns are connected underneath the board and the two long horizontal rows are connected and used as the power rails. It can become rather confusing if large circuits are modelled.
Matrix board	This type of board is made from an insulating material into which holes are drilled at regular intervals. Pins are pressed into the board where components need to be fixed in. Circuits can be laid out in an almost identical fashion to the circuit diagram, but this method does not work well for circuits containing ICs.
Veroboard/stripboard	Veroboard is very similar to matrix board with the addition of a series of copper tracks running down the underside. This allows components to be soldered directly to the board, which involves cutting tracks in certain places. Careful consideration must be given to the placing and orientation of ICs.
Printed circuit boards (PCBs)	PCBs are made from a copper-clad glass-reinforced plastic, which is an insulating material. The board undergoes a series of processes that involve transferring a circuit layout onto the copper either manually or photographically. The board is then etched in ferric chloride and drilled. Very complex circuits can be built up and any component can easily be placed into the circuit. Any mistakes, however, result in a new board having to be made.

EXAMINATION QUESTION

Example question and answer.

 Q1 *Give **one** advantage of using breadboard/ prototyping board for modelling circuit ideas.* **(2 marks)**

Acceptable answer
Breadboard does not involve permanently fixing components into place so **corrections can be easily made** by simply removing components and moving or replacing them.

PRACTICE EXAMINATION STYLE QUESTION

1 a) The motor drive system on a woodworking lathe utilises a 'cone pulley' mechanism. With the aid of annotated sketches, show what is meant by a 'cone pulley' mechanism. **(3)**

b) Explain what type of drive belt would be most suitable for this application and why. **(2)**

c) With the aid of a schematic diagram, show how it is possible to make the two pulleys in Figure 4.21 rotate in opposite directions. **(1)**

d) Explain **two** different methods of maintaining tension in belt drives. **(4)**

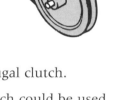

Fig. 4.21

e) Using notes and sketches, explain the principles of a centrifugal clutch. **(3)**

f) Give **one** example of an application where a centrifugal clutch could be used. **(1)**

g) Explain why garage mechanics need to consider health and safety issues when handling and replacing clutch materials. **(1)**

2 A coupling is a joint that is used to connect two rotating shafts together.

a) (i) Using annotated sketches, describe the principles of a coupling. **(5)**
(ii) Explain **one** advantage of using a flexible coupling rather than a flanged or split coupling. **(2)**

b) A cam is a specially shaped piece of metal or plastic.
(i) Draw a simple pear-shaped cam. Clearly label the key features of its profile or geometry. **(5)**
(ii) Figure 4.22 shows a particular type of cam being used as a bobbin winder. State the correct name of the cam being used in this application. **(1)**

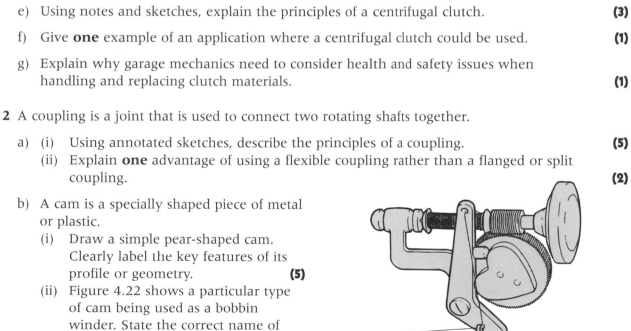

Fig. 4.22

(iii) Explain why this type of cam makes it ideal for this type of application. **(2)**

Total for this question paper: 30 marks

Design and technology capability (R6)

This is a three-hour design exam that focuses on the knowledge and understanding found within the designing and making process. As it is synoptic, this paper is taken at the end of the course and will examine all the areas covered during the two years of the course. The structure and content of the unit reflects the outline as detailed below.

You will receive the design research paper at least six weeks prior to the exam. This will set the context for the exam and will provide clear guidelines that will enable you to focus and target your research.

The exam is regarded as an 'open book' – whatever research you carry out and gather, you can take in with you and refer to in the exam. You should therefore give very careful consideration to what you research and to the amount of information you take in with you. If you take too much information, you could waste valuable time looking through it and trying to sift material out.

You should aim to organise your research information under the same headings that appear in the design specification criteria. However, research should be drawn from, and focused upon, the following areas:

- market research
- analysis of existing products
- research into materials, components and processes
- legal requirements and standards relating to quality and safety
- values issues, and moral and environmental issues, which might impact on the problem and solution.

Table 6.1 Mark allocation

Section heading	Marks available	Recommended time (mins)
a) Analyse the design problem and develop a product design specification, identifying appropriate constraints	15	30
b) Generate and evaluate a range of design ideas	15	30
c) Develop, describe and justify a final solution, identifying appropriate materials and components	15	30
d) Represent and illustrate your final solution	20	40
e) Draw up a production plan for your final solution	15	30
f) Evaluate your final solution against the product design specification and suggest improvements	10	20
Total	90	180

You will be provided with pre-printed standard A3 sheets in the exam. They have been designed to be all you should need. The pasting of pre-prepared or photocopied sheets is NOT permitted. ICT will NOT be assessed in this exam so you will not be allowed to use ICT facilities in the exam.

The pre-printed sheets are headed up as in Table 6.1, indicating both the marks available and the recommended amount of time to be spent.

Note: Always make sure that you read the instructions on the pre-printed sheets carefully. It is possible that at a future date the allocation of marks *could be slightly amended*, depending on the complexity of the product you are required to design.

Analyse the design problem and develop a specification

You need to

☐ use appropriate research techniques, primary sources and specialist information
☐ use appropriate analysis techniques to analyse the problem and identify purpose/users/target market
☐ use research analysis to develop a design specification to form the basis for generating and evaluating ideas.

KEY TERMS

Check you understand these terms

research analysis, design specification, constraints, attributes

Further information can be found in *Advanced Design and Technology for Edexcel, Product Design: Resistant Materials Technology*, **Unit 6, section a).**

Use appropriate research techniques, primary sources and specialist information

The problem must be thoroughly researched before you go into the exam. Your **research** can be drawn from the following areas:

• market research
• the Internet

• your course notes
• talking to experts
• CD-ROMs
• product **analysis**
• libraries
• your teachers.

Your research must be focused and relevant to the problem, but areas you should consider may include:

• purpose/function
• trends and fashions
• target market
• user and performance requirements
• the work of other designers
• relevant materials, components and systems
• relevant processes and technology
• legal and external standards/requirements (safety and quality)
• values issues
• cultural, moral, social and environmental issues.

a) Analyse the design problem and develop a product design specification, identifying appropriate constraints (15 marks)

Analyse the design problem (6 marks)

The analysis should help to clarify the problem. Your work should show that you understand the problem clearly. As part of your analysis you should:

- clearly identify and expand the areas and issues relevant to the problem that you have addressed in your research
- make reference to your research material.

It is not necessary to write at length – short and concise points are sufficient. Brainstorming, thought showers, mind mapping and tables are proven methods of organising this form of information.

Develop a product design specification, identifying appropriate constraints (9 marks)

The **design specification** should be generated from your research and analysis. You must indicate how you used your research in this section, such as where you found your information. The specification headings are printed at the top of your answer sheet, so use them as a guide. Your specification criteria/design requirements should be written as short, reasoned sentences. Your product design specification is very important because it forms the basis for generating and evaluating your design ideas and for evaluating your final design proposal.

Your specification criteria should be specific to the set design problem rather than points of a generic nature, with each of the points being justified. Most of the marks in this section are awarded for the justification of the points.

Constraints are design rules that provide limits for the design, such as 'no more than 1kg in weight'. **Attributes** are qualities that need to be incorporated into the design, such as 'portability'.

Your specification criteria should:

- be specific to the requirements of the set design problem
- be influenced by your research and analysis

- explain the design requirements of the product
- be measurable where possible.

The printed specification headings ask you to develop the following criteria:

- **purpose/function:** what the product and individual components are for and what they need to do
- **aesthetics:** how the product should look – its style, form and aesthetic characteristics
- **processes, technology and scale of production:** manufacturing processes and technology required to make the product at an appropriate level of production – either one-off, batch or mass production
- **cultural, social, moral and environmental issues:** these may influence your design ideas
- **market and user requirements:** market trends, user needs and preferences, ergonomic constraints
- **quality control:** the quality requirements of your product (and target market) and how you will achieve quality using quality control and quality standards.

Examiner's Tips

- It is very important to refer to the research material that you have generated. The examiner needs to see that you have done the research and are making use of it.
- Make sure that you:
  - expand on your initial points
  - link your points to the problem
  - use your research
  - always justify points made.
- You must concentrate a good proportion of your research on the construction details of products within the design context. You will need to find out about the specific materials and processes used. Remember that you will have to design something similar in the exam.

Generate ideas; evaluate each idea and justify decisions made

You need to

☐ **generate and record a range of design ideas using appropriate communication techniques, such as graphical/numerical/language systems**
☐ **evaluate ideas against the design specification to determine their feasibility**
☐ **justify the design selected for development.**

KEY TERMS
Check you understand these terms

annotations, evaluate, justify

Further information can be found in *Advanced Design and Technology for Edexcel, Product Design: Resistant Materials Technology*, Unit 6, section b).

b) Generate and evaluate a range of design ideas. Use appropriate communication techniques and justify decisions made (15 marks)

Generate and record a range of design ideas (10 marks)

You are expected to generate a minimum of three realistic, feasible ideas in response to the main problem. Your initial ideas should:

• be clear sketches
• provide clear detailed **annotations** (notes)
• make reference to the scale of production and commercial processes
• make reference to specific materials and processes.

Annotation means supplementing your design ideas with notes giving details about material, processes, manufacturing techniques and detail. Any annotation must be specific to your chosen materials rather than generic detail, such as 'wood' or 'plastic'. Details regarding manufacturing techniques and processes should reflect specifics too – this is a good opportunity to make use of your research information and to show what you have learned over the two-year course.

Evaluate design ideas and justify decisions made (5 marks)

Once you have completed your ideas, you should **evaluate** them against your product design specification. This can be achieved through:

• annotation, which refers clearly to individual specification points
• tables, which assess each idea against the specification points.

At the very least, you should **justify** each of your ideas by providing advantages and disadvantages. The examiners are asked to look at the following areas:

• a minimum of three ideas
• material details
• methods of production
• quality of communication
• evaluation/justification.

Examiner's Tip

The task outlined on the first page of the exam paper will often provide you with one or more constraints that need to be included in your analysis and specification. Read the task carefully and highlight any constraints before you start.

Develop the chosen idea into an optimum solution; describe and justify a solution

You need to

☐ **describe and justify the solution in terms of function, appearance, characteristics, materials, components/systems, processes and technological features to be used.**

KEY TERMS

Check you
understand this term

development

Further information can be found in *Advanced Design and Technology for Edexcel, Product Design: Resistant Materials Technology*, Unit 6, section c).

c) Develop, describe and justify a final solution, identifying appropriate materials and components (15 marks)

In this section you are asked to develop and refine your initial design ideas. Both 2D and 3D sketches can be used, but there must be obvious changes from the initial ideas section. There must be evidence of how you have improved and changed aspects/features in an attempt to improve and better your product design.

Develop a final solution, identifying appropriate materials and components (8 marks)

Annotation should be clear and concise. Details should be sufficient to show and explain how the product will work and function according to the product specification. Examiners are looking for:

* evidence of **development**
* does it work?
* are the materials suitable?
* methods of production/manufacture.

Describe and justify your final solution (7 marks)

In the second part of this section, you are asked to describe and justify your final solution. This should be an account explaining why and how your developed design is the best solution to the problem. You are given a series of headings to guide you on the pre-printed sheet. Your description and justification of your design solution should be made against the following criteria:

* function
* appearance
* performance
* materials
* components/systems
* processes
* technological features.

Examiner's Tips

* The examiner needs to follow your train of thought. It is a good idea to include some arrows so that your developing ideas can be logically followed through.
* You may separate your sheets carefully if you wish. This will allow you to refer more easily to your specification when you need to. You must, however, ensure that all of your work is reassembled in the correct order using a treasury tag at the end of the exam. It should not be necessary to add sheets.

Represent and illustrate the final solution

You need to

☐ represent and illustrate the final solution using clear and appropriate communication techniques.

KEY TERMS

Check you understand these terms

final design solution, cutting list

  Further information can be found in *Advanced Design and Technology for Edexcel, Product Design: Resistant Materials Technology*, **Unit 6, section d).**

d) Represent and illustrate your final solution (20 marks)

You need to represent and illustrate your final design solution, using

- clear manufacturing and assembly details (at least eight)   (8 marks)
- dimensions/sizes (at least eight)   (4 marks)
- details and quantity of materials/components   (4 marks)
- clear and appropriate communication techniques.   (4 marks)

In this section, you are asked to 'represent and illustrate' your **final design solution**. This does not mean you redraw what you presented in the previous section. You should produce a drawing of your final proposed solution and give a clear indication of dimensions, construction details, quantities involved and

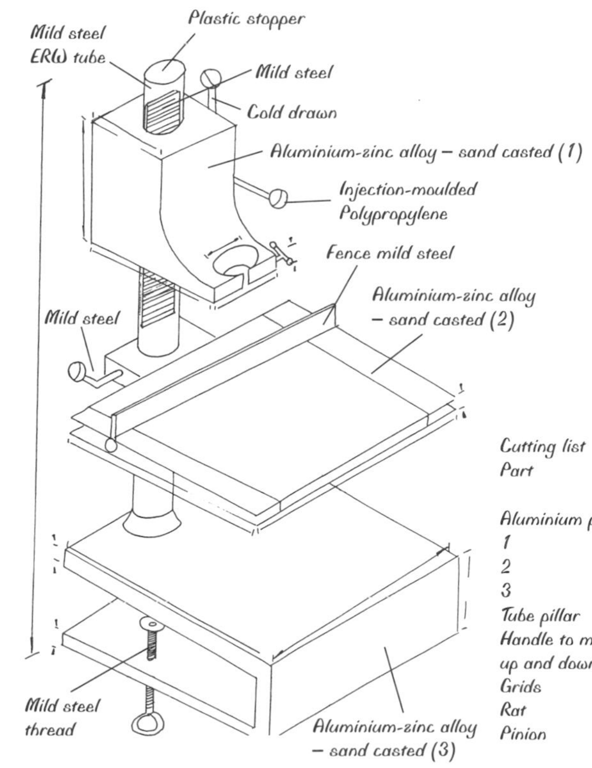

Mild steel ERW tube
Plastic stopper
Mild steel
Cold drawn
Aluminium-zinc alloy – sand casted (1)
Injection-moulded Polypropylene
Fence mild steel
Aluminium-zinc alloy – sand casted (2)
Mild steel
Mild steel thread
Aluminium-zinc alloy – sand casted (3)

Cutting list

Part	Material	Length	Width	Depth variable	Quantity
Aluminium part					
1	Aluminium-zinc alloy	110mm	115mm		1
2	Aluminium-zinc alloy	200mm	250mm	15mm	1
3	Aluminium-zinc alloy	200mm	250mm	Variable	1
Tube pillar	Mild steel tubing	470mm	40mm	-	1
Handle to move up and down	Mild steel rod	100mm	10mm	-	3
Grids	Polypropylene	-	-	-	4
Rat	Mild steel	250mm	35mm	5mm	1
Pinion	Mild steel	100mm	15mm	-	1

Other orderable parts:
Plastic stoppers – 1 – for end of pillar
Screw thread x 2 with handle on end
Screws and bolts
Aluminium alloy – TIG welded to other parts

Fig. 6.1 A student's representation and illustration of the final solution

details about materials and components. Sufficient detail should be provided for the examiner to see and understand how your final proposal is to be manufactured on an appropriate scale of production.

Draw up a production plan

You need to

describe the production requirements of the solution, to include where appropriate:

☐ **assembly processes/unit operations**
☐ **sequence of assembly/work order with details of tools, equipment, tolerances**
☐ **quality checks at critical control points, with quality indicators.**

KEY TERMS

Check you understand this term

critical control points

Further information can be found in *Advanced Design and Technology for Edexcel, Product Design: Resistant Materials Technology*, Unit 6, section e).

e) Draw up a production plan for your final solution (15 marks)

There is more than one way of producing a production plan. One proven method of incorporating all the information required is to

A variety of graphical techniques could be used, including dimensioned drawings, sections and cut-away sections, exploded views and hidden detail. A **cutting list** and parts list should be provided that shows all the details and quantities of the materials and components used. Marks are awarded in this section for the use of clear and appropriate communication techniques.

Examiners will be looking for:

• details about manufacture
• assembly
• dimensions
• quantities
• appropriate levels of communication.

produce a simple flow chart. Make sure you address all the elements of the task.

Your plan should include operations related to all of the following:

• processing (e.g. injection moulding, casting)
• assembly (e.g. screw-on tops, nuts/bolts,)
• specific quality control checks (how, when and why).

Assembly processes/unit operations

Your plan must incorporate the manufacture of all the component parts that go together to make up your product. You must identify the specific operations and commercial processes that will be used. Do not include:

• the design stages
• any modelling processes.

The nature of the processing and assembly required will depend upon your chosen materials and the scale of production identified in your product design specification. Make sure that you use specific and appropriate technical terms.

Sequence of assembly/work order with details of tools, equipment and tolerances

Once you have identified your processes, you need to organise them into the correct sequence. Simple flow charts are a proven method of representing the sequence of operations. You should refer to specific tools and equipment. You will address tolerances in your quality control checks.

Quality checks at critical control points with quality indicators

You should identify **critical control points** where quality checks will be made. These should be specific; you need to identify:

- where (at what point) in the process the quality checks will be made
- what is being checked (critical dimensions against tolerances)

- how it is being checked (e.g. visual inspection, manual gauge or micrometer).

Do not use generalised statements, such as 'the component is checked for quality'. *You need to state exactly where and how you will check for quality – this must be clearly identified on the production plan.*

Examiner's Tip

Be specific. You will not gain credit for generalised statements. Address all of the areas as equally as possible for all components. Ideally, you should be able to describe a minimum of five processing operations, five assembly operations and five quality control procedures.

Evaluate

You need to

☐ **evaluate your final solution against the product design specification.**

KEY TERMS

Check you understand this term

evaluation

Further information can be found in *Advanced Design and Technology Edexcel, Product Design: Resistant Materials Technology*, Unit 6, section f).

Evaluate your final solution against the product design specification and suggest improvements (10 marks)

Your **evaluation** and your suggestions for improvement are equally important, so divide your time accordingly. Make sure you address all the elements of the task and refer to your specification. One mark is awarded for each valid, justified point.

Evaluate your final solution against the product specification (5 marks)

Your evaluative comments should be:

- objective (positive and negative points relating closely to specification)
- justified (explained with reasons given to support points made).

If you use the specification, you will be addressing appropriate issues by using the headings given in the first exam question. You should be critical: by identifying negative

Examiner's Tips

- The marks for this section are usually divided equally between your evaluation and suggested modifications. Make sure that you provide at least five justified points for each. If you do not make any points, you will not receive any marks.

Two things are essential to achieving a good mark in this unit:

- thorough preparation before the exam: use the preparation sheet to prepare and practise
- good time management during the exam: stick to the times given on the exam paper.

aspects of your design, you will find it easier to suggest improvements. Turn these headings into questions: for example, will your product perform all the functions listed in your specification?

Suggest improvements (5 marks)

You should suggest realistic improvements that would enable your product to better meet the needs of your product design specification. Generalised, unjustified points will not be credited, nor will comments related to the quality of your work. Your suggested improvements should relate to the success of your final design proposals. Again, you need to refer to your specification. Consider, for example:

- suggesting modifications to improve the aesthetics, ergonomics or durability of your product
- suggesting changes to your design to optimise manufacturing processes
- changing your choice of materials or manufacturing processes
- recommending more use of CAM or automation for the manufacture of your product.

PRACTICE EXAMINATION STYLE QUESTION

Gardeners often need a means of holding rubbish bags open.

You have investigated:

- portable garden tools for waste handling
- various types of garden waste
- sizes of plastic sacks.

Your task is to design a holding system to keep open a plastic bag whilst it is filled up with grass and leaves.

(i)   the system must be collapsible when not in use
(ii) allow for the bag to be easily transported once full
(iii) keep the bag wide open whilst it is being filled.

   a) Analyse the design problem and draw up a product design specification, identifying appropriate constraints. (1

   b) Generate and evaluate a range of design ideas. (1

   c) Develop, describe and justify a final solution, identifying appropriate materials and components. (

   d) Represent and illustrate your final solution. (2

   e) Draw up a production plan for your final solution. (

   f) Evaluate your final solution against the product design specification and suggest improvements. (10)

Total for this question paper: 90 marks